To my wife, Elena,
and our children,
Iván, Leo, Elena, and Hugo.

Preface

This book is the first volume in the *QuantPhi Series in Mathematical Finance*, a collection of books dedicated to presenting, in a concise, detailed, and rigorous manner, the stochastic foundations of quantitative finance. This series will include volumes on option pricing theory, the Black-Scholes model, its extensions, and interest rate models.

This volume, as well as the other books in the series, has been written in accordance with the QuantPhi principles.[1] In particular, it has been conceived and written following the traditional approach of academic authorship. The mathematical content, explanations, proofs, examples, and exercises are the result of careful study, synthesis and exposition by the author. Generative artificial intelligence has been used only selectively and transparently, mainly for the creation of illustrations, non-technical material, and the proofreading of the final version of the manuscript, while the technical and educational content has been developed entirely through human intellectual work. Another principle followed throughout the book is that mathematical results should not be presented without proof. With the exception of statements whose proofs are truly immediate, every result is either accompanied by a complete proof or by an explicit reference where a detailed proof can be found.

Why this book?

A few years ago, I published a book entitled "The Mathematics of Financial Markets Within Everyone's Reach: An Introduction Without Formulas in the Main Text". The aim of the book was to provide anyone interested in quantitative finance with a first introduction to the underlying mathematical models, through a qualitative and accessible presentation, without entering into formal developments. After publishing this "gentle introduction" to quantitative finance, it seemed natural to follow it with a formal and mathematically rigorous treatment, intended for readers who, having read the first book and possessing the necessary mathematical background, wished to take the leap into mathematical finance.

Shortly after beginning this project, I realized that, if it was to be self-contained, it would become a very long book. Covering stochastic calculus, arbitrage theory, market completeness, option pricing, the Black-Scholes model and its extensions, local and stochastic volatility models, as well as interest rate models and their derivatives within a single volume, would have resulted in a book of considerable length, likely discouraging readers approaching the subject for the first time.

[1] The complete text of the principles can be consulted at `https://quant-phi.com/our-principles`.

Alberto Bueno Guerrero

Introduction to Stochastic Calculus

QuantPhi Series in Mathematical Finance

Volume I

QΦ
QuantPhi

Published by QuantPhi

First Edition, 2026

ISBN: 979-13-992628-0-3

Printed independently through Amazon Kindle Direct Publishing (KDP).

For updates, supplementary material, and additional resources, please visit
`https://quant-phi.com`

For this reason, together with QuantPhi, I decided to divide this material into a series of volumes that constitute the *QuantPhi Series in Mathematical Finance*, the first volume of which you now hold in your hands. Although each volume is self-contained, together they provide a coherent and rigorous introduction to the mathematical methods that underpin modern quantitative finance.

What is this book about?

This book is intended to be a self-contained introduction to stochastic calculus for Itô processes, providing readers with the mathematical tools needed to study many of the continuous-time mathematical models used in quantitative finance, which will be developed in subsequent volumes of this collection.

Beyond introducing the subject, I believe the book can be used in two main ways. The first is as a classic textbook for any university-level course with mathematical content. Readers interested in using it in this way will find everything they need: an organized and detailed theoretical development, with most proofs included, as well as worked examples and solved exercises and problems, that will allow them to develop and test their understanding. The index at the end of the book will allow readers to easily locate the definitions and results they need to consult during their studies. Furthermore, readers who prefer a more traditional lecture format may also benefit from the pre-recorded video lectures of the *QuantPhi Program in Mathematical Finance*, available through the QuantPhi website.[2]

The second use of the book is as a reference text. One of the problems commonly faced by students or researchers in quantitative finance, as well as quantitative analysts, is the difficulty in finding a reliable source to clarify a particular point, recall the assumptions of a model, or verify the conditions under which it is valid. My own experience tells me that it is often necessary to consult several textbooks, and even then, sometimes the proof or argument you need is not found in the available sources.

While it is clearly impossible to encompass all the knowledge related to a particular subject in a single volume, with this book, I have attempted to provide the reader with a comprehensive collection of definitions, results, and proofs related to this field of knowledge, and especially those related to quantitative finance. To this end, I have used a considerable number of academic references, which can be consulted at the end of the book.

Therefore, if you find yourself in any of the situations mentioned above, and you have this book on your desk, whenever you need to recall a concept or a particular result in stochastic calculus, you only need to open the book to the index to locate the definition of the concept or the statement of a particular result and, in most cases, its proof.

To reinforce these two uses of the book—as a textbook and as a reference text—I have included two mathematical appendices with virtually all the definitions and results of functional analysis and probability theory necessary to study the material covered in the book.

How is this book organized?

The book comprises four chapters, two mathematical appendices, a bibliography, and an index. The first three chapters are dedicated, in this order, to Brownian motion,

[2] `https://quant-phi.com`

stochastic integration, and stochastic differential equations. The fourth chapter is devoted to the solutions to the problems proposed in each of the first three chapters. The first of the two appendices presents content related to measure theory, integration, and functional analysis, while the second is dedicated to probability theory.

Each chapter begins with an introduction that presents the topics to be covered. The theoretical content is then developed. In each definition, the term being defined appears in bold in the heading, making it easier for the reader to locate. Mathematical results, whether corollaries, lemmas, propositions, or theorems, include their complete proofs, with the exception of a few that have been omitted due to their length or excessive technical complexity for the level of this book. In all cases where the proof has been omitted, one or more references have been provided where the reader can find it. Readers approaching the subject for the first time may wish to skip the more technical proofs on a first reading and return to them later.

Interspersed throughout the theoretical development are examples, as well as proposed exercises related to the previous content, allowing the reader to test their understanding. All exercises are solved in detail, and their solutions appear in a section immediately following the theoretical content of the corresponding chapter. After the solutions to the exercises in each chapter, the reader will find a list of problems, slightly more difficult than the exercises and related to the chapter's content, which will allow the reader to consolidate the material covered in the chapter. The solutions to the problems in the first three chapters can be found in the fourth chapter, which is dedicated exclusively to them.

Each chapter ends with a section called "Notes," which provides information about the main references used in preparing the chapter, the decisions made regarding the content presented, and additional related information.

The appendices present, in a concise and schematic way, the main definitions and mathematical results required in the book, and some additional ones that will be useful to readers who continue their studies in the field of mathematical finance. Although the appendices cover a broad range of topics, they should not be considered a miniature course on the concepts covered, but rather a quick reference guide. Therefore, unlike the main content of the book, most proofs are not explicitly stated, although references to where they can be found are provided.

The first appendix is entitled "Measure Theory, Integration and Functional Analysis" and contains the basic concepts of these disciplines necessary for some proofs and, mostly, to provide a mathematical basis for what is presented in the second appendix, entitled "Probability Theory," which provides the general framework in which stochastic calculus is developed.

Finally, an alphabetical index at the end of the book enables readers to locate definitions, results, and concepts quickly, reinforcing the role of the book as a reference text.

What prior knowledge is required?

Although I have tried to make the book self-contained, and many of the necessary mathematical tools appear in the appendices, a certain level of mathematical knowledge is required to get the most out of the book.

Specifically, it is highly recommended to have knowledge of single- and multivariable calculus at the level of an introductory university course, basic methods for solving ordinary differential equations, and the basic concepts of partial differential equations. As far as linear algebra is concerned, a basic familiarity with matrices is sufficient. Although probability theory is developed in an appendix, it is nevertheless helpful for readers to have an introductory knowledge of probability, allowing them to approach the formal theory from an intuitive perspective.

With this mathematical background, readers should be able to embark confidently on the study of stochastic calculus and, when necessary, consult the appendices for any unfamiliar mathematical concepts and results.

Nevertheless, motivated readers who do not yet possess all of these prerequisites should not feel discouraged. Those with at least a pre-university mathematical background may still benefit from the book, provided they are willing to learn, as needed, the additional concepts and techniques encountered along the way.

Who is this book for?

This book is conceived as the first volume in a collection dedicated to mathematical modeling in finance. Therefore, a natural readership would include university students or graduates with STEM degrees, such as physicists, mathematicians, engineers, economists and statisticians, who wish to pursue a career in quantitative finance or simply explore this fascinating field.

Furthermore, since the quantitative finance models themselves are deliberately postponed to subsequent volumes, this book contains virtually no financial applications of the mathematical tools it develops, apart from a few isolated references. For all these reasons, the book will be of interest to readers with a mathematical background who, while not initially interested in mathematical finance, wish to study a branch of mathematics that they are unlikely to have encountered during their undergraduate studies.

The book may also serve as recommended reading for postgraduate courses in stochastic calculus. As I mentioned earlier, quantitative analysts and researchers in quantitative finance will find this book a useful reference text, both for refreshing forgotten concepts and for acquiring new knowledge that was not covered during their formal education.

Finally, anyone who, even if unfamiliar with the financial world, is curious and in need of intellectual stimulation, may find in this book an engaging intellectual challenge that could lead them to the exciting world of quantitative finance.

I hope that, whether you identify with any of the groups described above or not, you will enjoy reading and studying this book and find the experience truly rewarding.

Granada
July 2026

Alberto Bueno Guerrero

Contents

Chapter 1

Brownian Motion

Brownian motion is a key stochastic process in the mathematical modelling of financial markets. Most of the processes used to describe financial asset prices are based on it in one way or another. Brownian motion is named after the Scottish botanist Robert Brown, who discovered it in 1827 while observing the movement of pollen grains suspended in water. Later, in one of his famous 1905 papers, Albert Einstein used kinetic molecular theory to explain Brownian motion, thereby providing support for the atomistic hypothesis. The mathematical formulation of the concept was later developed by Norbert Wiener, who established the defining properties of Brownian motion, which consequently also became known as the Wiener process.

In this chapter we study the mathematical theory of Brownian motion in sufficient depth to provide a solid foundation for the concepts of stochastic integral and stochastic differential equation, which will be developed in the following chapters.

We begin in Section 1.1 with a brief description of Einstein's approach to the problem. Section 1.2 introduces the concept of stochastic process and presents adapted processes and martingales. The mathematical definition of Brownian motion, together with its first properties, is given in Section 1.3. Section 1.4 is devoted to martingales derived from Brownian motion.

In Section 1.5 we define Gaussian processes and prove that Brownian motion belongs to this class. Several transformations of Brownian motion that preserve the Brownian property are studied in Section 1.6. Some of the most remarkable properties of Brownian motion concern the behaviour of its sample paths; these are studied in Section 1.7. The memoryless property of Brownian motion, namely the Markov property, together with its strong version, is presented in Section 1.8. Applications of the strong Markov property to the study of hitting times are discussed in Section 1.9. Hitting times are then applied to the study of the running maximum of Brownian motion in Section 1.10. The extension of Brownian motion to higher dimensions is developed in Section 1.11. The chapter concludes with the solutions to the proposed exercises, a collection of supplementary problems, and the notes, presented in Sections 1.12, 1.13, and 1.14, respectively.

Sections 1.1 and 1.7, marked with asterisks, are not strictly necessary for the study of stochastic calculus and may therefore be omitted.

1.1 The Physical Approach to Brownian Motion*

In his 1905 paper, Einstein carried out a probabilistic study of the one-dimensional motion of an ensemble of particles suspended in a liquid. The assumptions adopted by Einstein were the following:

i) The displacements of a particle in two consecutive and small time intervals of magnitude τ are mutually independent.

ii) The displacement Δ of the particle in the time interval τ is a continuous random variable, with density function $\phi(\Delta)$.

iii) Collisions between the fluid molecules and the particles occur with the same probability in any direction, which leads to the symmetry condition

$$\phi(-\Delta) = \phi(\Delta)$$

The number of particles in the interval $(x, x + dx)$ at time t is given by $n(x,t)\,dx$, where $n(x,t)$ is the particle density at point x at time t. By conservation of the number of particles, the number of them in the interval $(x, x + dx)$ at the later time $t + \tau$, namely $n(x, t + \tau)\,dx$, is obtained as follows. The number of particles in the interval $(x - \Delta, x - \Delta + dx)$ at time t, is $n(x - \Delta, t)\,dx$, while the probability that a particle moves from the interval $(x - \Delta, x - \Delta + dx)$ to the interval $(x, x + dx)$ in the time interval τ is, by assumptions i) and ii), equal to $\phi(\Delta)$. Integrating over all possible values of Δ, we obtain

$$n(x, t + \tau)\,dx = \int_{\Delta=-\infty}^{+\infty} n(x - \Delta, t)\,dx\phi(\Delta)\,d\Delta \tag{1.1}$$

On the other hand, expanding in Taylor series and considering that both τ and Δ are small, we can approximate $n(x, t + \tau)$ and $n(x - \Delta, t)$ as

$$n(x, t + \tau) \simeq n(x,t) + \tau\frac{\partial n(x,t)}{\partial t}$$
$$n(x - \Delta, t) \simeq n(x,t) - \Delta\frac{\partial n(x,t)}{\partial x} + \frac{\Delta^2}{2}\frac{\partial^2 n(x,t)}{\partial x^2}$$

Substituting these expressions into (1.1), we obtain

$$\begin{aligned} n(x,t) + \tau\frac{\partial n(x,t)}{\partial t} &= n(x,t)\int_{-\infty}^{+\infty}\phi(\Delta)\,d\Delta - \frac{\partial n(x,t)}{\partial x}\int_{-\infty}^{+\infty}\Delta\phi(\Delta)\,d\Delta \\ &\quad + \frac{1}{2}\frac{\partial^2 n(x,t)}{\partial x^2}\int_{-\infty}^{+\infty}\Delta^2\phi(\Delta)\,d\Delta \\ &= n(x,t) + \frac{1}{2}\frac{\partial^2 n(x,t)}{\partial x^2}\int_{-\infty}^{+\infty}\Delta^2\phi(\Delta)\,d\Delta \end{aligned}$$

where we have applied that $\int_{-\infty}^{+\infty}\phi(\Delta)\,d\Delta = 1$, and that $\int_{-\infty}^{+\infty}\Delta\phi(\Delta)\,d\Delta = 0$ by assumption iii). Simplifying the previous equation, we arrive at

$$\frac{\partial n(x,t)}{\partial t} = D\frac{\partial^2 n(x,t)}{\partial x^2} \tag{1.2}$$

where we have defined $D = (2\tau)^{-1} \int_{-\infty}^{+\infty} \Delta^2 \phi(\Delta)\, d\Delta$. Equation (1.2) is a parabolic partial differential equation (PDE) that determines the time evolution of the particle density field $n(x,t)$. In Physics and Engineering, this equation is known as the one-dimensional *heat equation*, and it describes the heat flow in an homogeneous and isotropic medium with thermal diffusivity D, where $n(x,t)$ represents the temperature at point x of the medium at time t.

Equation (1.2) can be solved analytically in the case in which all particles are initially located at the origin, that is, $n(x,0) = \delta(x)$, where δ is the Dirac delta. Under this assumption the solution is given by

$$n(x,t) = \frac{1}{\sqrt{4\pi Dt}} e^{-\frac{x^2}{4Dt}} \tag{1.3}$$

which is the probability density function of a Gaussian random variable with mean zero and variance $2Dt$.

1.2 Stochastic Processes

A stochastic process describes the evolution of a random variable over time. The theory of stochastic processes is tailor-made for the study of financial markets, since they determine the change over time in the prices of financial assets, which can be considered as random variables.

We will assume the existence of a filtered probability space $\left(\Omega, \mathcal{F}, (\mathcal{F}_t)_{t\geq 0}, \mathbb{P}\right)$, with $(\mathcal{F}_t)_{t\geq 0}$ the filtration satisfying the usual hypotheses. In financial applications, the filtration is interpreted as representing the temporal flow of information, that is, $\mathcal{F}_t$ contains all the information available in the market at time t. The increasing character of filtrations, i.e., $\mathcal{F}_s \subset \mathcal{F}_t$ for $s < t$ ensures that information accumulates over time.

Moreover, throughout this book, unless otherwise stated, we shall assume that the initial σ-algebra is the completion of the trivial σ-algebra $\{\varnothing, \Omega\}$, namely

$$\mathcal{F}_0 = \{A \in \mathcal{F} : \mathbb{P}(A) \in \{0,1\}\}.$$

Since every $\mathcal{F}_0$-measurable random variable is constant almost surely, Proposition B.62 yields that, for every integrable random variable X,

$$\mathbb{E}[X|\mathcal{F}_0] = \mathbb{E}[X] \text{ a.s.}$$

Definition 1.1. (Stochastic process) A (continuous time) stochastic process X is a collection $X = \{X_t\}_{t\geq 0}$ of random variables on $(\Omega, \mathcal{F})$.

Since a random variable is itself a (measurable) function from Ω to $\mathbb{R}$, a stochastic process X can be considered as a function of two variables

$$\begin{aligned} X: \quad \Omega \times [0,\infty) &\longrightarrow \mathbb{R} \\ X(\omega,t) &\longmapsto X_t(\omega) \end{aligned}$$

For each $t \in [0,\infty)$, the mapping $\omega \mapsto X_t(\omega)$ is a random variable, and for each $\omega \in \Omega$, the function $t \mapsto X_t(\omega)$ is a *sample path*, *trajectory*, or *realization* of the stochastic process.

It is important to distinguish between a stochastic process and each of its sample paths. If we start a random experiment and record its results (random variables) over time, we are observing one of the possible sample paths (for a specific $\omega_1 \in \Omega$) associated with a stochastic process. If we restart the experiment, we will observe a different sample path (for another $\omega_2 \in \Omega$). The stochastic process, which is related to the experiment itself, refers to the set of all possible realizations or sample paths. In the context of financial markets, when we look at a chart with the past prices of an asset, we are observing a sample path, for a given $\omega \in \Omega$, of the stochastic process associated with that asset.

In most of the applications, it is important to know whether the value of a stochastic process at any given time is included in the information available up to that time. When a stochastic process satisfies this property we will say that it is an adapted process. The mathematical definition is as follows.

Definition 1.2. (Adapted process) The stochastic process X is adapted if X_t is an $\mathcal{F}_t$-measurable random variable for each $t \geq 0$.

Sometimes, given a stochastic process X, it is useful to consider the filtration generated by the process itself

$$\mathcal{F}_t^X = \sigma\left(X_s : 0 \leq s \leq t\right)$$

which is the smallest σ-field with respect to which X_s is measurable for every $s \in [0, t]$. This filtration is called the *natural filtration* of the process X. Obviously, every process is adapted when the filtration of the probability space is its natural filtration.

The following definition introduces a class of adapted processes that plays a crucial role in mathematical finance, especially in arbitrage theory and option pricing.

Definition 1.3. (Martingale) A martingale is an adapted process X satisfying

i) $\mathbb{E}\left[|X_t|\right] < \infty$ for every $t \geq 0$

ii) $\mathbb{E}\left[X_t \,|\mathcal{F}_s\right] = X_s$ a.s. for every s, t with $0 \leq s \leq t$

Condition i) is an integrability condition, while condition ii) states that the expectation of any future value of the process, given the information currently available, is equal to the current value of the process. In other words, of all possible future paths from the current time, we expect that the positive deviations balance the negative ones on average. This interpretation allows martingales to be interpreted as fair games, in which there is no systematic drift towards profits or losses. As an immediate consequence of Definition 1.3, we have that for a martingale X it is satisfied that

$$\mathbb{E}\left[X_t - X_s \,|\mathcal{F}_s\right] = 0$$

for every s, t with $0 \leq s \leq t$, i.e., conditional on the available information, the expected value of the change of a martingale is zero.

We finish this section dedicated to stochastic processes with the definition of Gaussian processes.

Definition 1.4. (Gaussian process) The process X is Gaussian if, for any integer $k \geq 1$ and real numbers $t_1, \ldots, t_k$, with $0 \leq t_1 < \cdots < t_k$, the random vector $\left(X_{t_1}, \ldots, X_{t_k}\right)^\top$ has a multivariate Gaussian distribution.

By Definition 1.4, the finite-dimensional distributions of a Gaussian process are completely determined by its mean and covariance functions $\mu(t) = \mathbb{E}[X_t]$ and $c(s, t) = \mathrm{Cov}(X_s, X_t) = \mathbb{E}\left[(X_s - \mu(t))(X_t - \mu(t))\right]$.

1.3 Definition and First Properties of Brownian Motion

We are now in a position to define what is perhaps the most important continuous-time stochastic process.

Definition 1.5. (Brownian motion) A stochastic process W is a Brownian motion if it satisfies

i) (Adapted) W_t is an $\mathcal{F}_t$-measurable random variable for each $t \geq 0$.

ii) (Increments independent of the past) For $0 \leq s < t$ the increment $W_t - W_s$ is independent of $\mathcal{F}_s$.

iii) (Gaussian increments) For $0 \leq s < t$ the increment $W_t - W_s$ is normally distributed with mean zero and variance $t - s$.

iv) (Continuous paths) The function $t \mapsto W_t(\omega)$ is continuous for almost every $\omega \in \Omega$.

If $W_0 = x$ a.s., the process W is called *Brownian motion started at* x. When $x = 0$, W is a *standard Brownian motion*. From now on, we will consider Brownian motion to be standard unless otherwise specified. Figure 1.1 shows two sample paths of a Brownian motion.

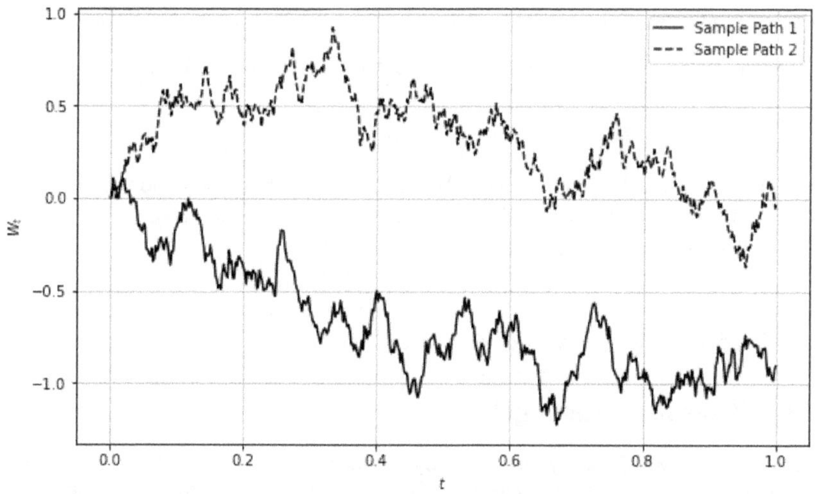

Figure 1.1: Two sample paths of Brownian motion.

Remark 1.6. There is an alternative definition of Brownian motion in which it is not necessary to consider a filtration (see, for example, Shreve (2004) or Klebaner (2005)). Under this definition, the process is not required to be adapted, and the condition of increments independent of the past is replaced by that of independent increments.[1]

[1] A stochastic process has *independent increments* if $X_{t_1} - X_{t_0}, \ldots, X_{t_n} - X_{t_{n-1}}$ are independent random variables, for every integer $n \geq 1$ and indices $t_0 < t_1 < \cdots < t_n$.

The Brownian motion defined in this way satisfies Definition 1.5 if we consider the filtration $\mathcal{F}_t^W$ generated by the process itself, since in this way the process is adapted and the increments are independent of the past (Karatzas and Shreve (1991), problem 1.4 of Chapter 2).

Remark 1.7. The continuity of paths is not actually independent of the other conditions since the Gaussian structure of increments guarantees the existence of a continuous modification of the process (Klebaner, 2005).[2]

Remark 1.8. The first question after defining Brownian motion is whether there really exists a process satisfying conditions i)-iv) in Definition 1.5. The answer is affirmative, and there are several constructions of Brownian motion, but all of them are outside the scope of this book. In Karatzas and Shreve (1991) the reader can find three different ways of obtaining explicitly a Brownian motion, as well as references to the original works.

Remark 1.9. As a consequence of Definition 1.5, Brownian motion satisfies the independent increments property. Moreover, the distribution of $W_t - W_s$ depends on t and s only through the difference $t - s$, a property known as *stationary increments*.

The following result is a direct consequence of Definition 1.5.

Proposition 1.10. *If W is a Brownian motion, then, for all $t > 0$, the random variable W_t has a Gaussian distribution with $\mathbb{E}[W_t] = 0$ and $\text{Var}[W_t] = \mathbb{E}[W_t^2] = t$.*

Proof. Immediate by considering that $W_t = W_t - W_0$, and iii) of Definition 1.5. □

From Proposition 1.10, we can write the density function of W_t as

$$\rho_t(x) = \frac{1}{\sqrt{2\pi t}} e^{-\frac{x^2}{2t}} \tag{1.4}$$

which coincides with that obtained using the physical approach, equation (1.3), making $D = \frac{1}{2}$. In other words, the density function of Brownian motion is the solution of the heat equation

$$\frac{\partial \rho_t(x)}{\partial t} - \frac{1}{2}\frac{\partial^2 \rho_t(x)}{\partial x^2} = 0$$

with initial condition $\rho_0(x) = \delta(x)$.

1.4 Brownian Martingales

In this section we will study three martingales related to Brownian motion, which will be important in what follows. The first of these martingales is Brownian motion itself.

Theorem 1.11. *Brownian motion is a martingale.*

Proof. Let W be a Brownian motion. By i) of Definition 1.5, W is an adapted process. From (1.4) we have

$$\mathbb{E}[|W_t|] = \frac{1}{\sqrt{2\pi t}} \int_{-\infty}^{+\infty} |x| \, e^{-\frac{x^2}{2t}} dx = \sqrt{\frac{2t}{\pi}} < \infty \quad \forall t \geq 0$$

[2] The stochastic processes X and Y are called *modifications* of one another if $\mathbb{P}[X_t = Y_t] = 1$ for all $t \geq 0$.

which proves i) of Definition 1.3. For condition ii), we have, for $0 \leq s \leq t$

$$\begin{aligned}\mathbb{E}\left[W_t \,|\mathcal{F}_s\right] &= \mathbb{E}\left[W_t - W_s + W_s \,|\mathcal{F}_s\right] = \mathbb{E}\left[W_t - W_s \,|\mathcal{F}_s\right] + \mathbb{E}\left[W_s \,|\mathcal{F}_s\right] \\ &= \mathbb{E}\left[W_t - W_s\right] + W_s = W_s\end{aligned}$$

where we have used i)-iii) of Definition 1.5 and the properties of conditional expectations. □

In addition to Brownian motion, there are other martingales derived from it that are interesting to know.

Proposition 1.12. *If W_t is a Brownian motion, the processes $W_t^2 - t$ and $e^{uW_t - \frac{1}{2}u^2 t}$ are martingales.*

Proof. Both processes are adapted by their definition. For the first one, we have

$$\mathbb{E}\left[\left|W_t^2 - t\right|\right] \leq \mathbb{E}\left[W_t^2 + t\right] = 2t < \infty$$

On the other hand

$$\begin{aligned}\mathbb{E}\left[W_t^2 \,|\mathcal{F}_s\right] &= \mathbb{E}\left[(W_t - W_s + W_s)^2 \,|\mathcal{F}_s\right] = \mathbb{E}\left[(W_t - W_s)^2 \,|\mathcal{F}_s\right] + \mathbb{E}\left[W_s^2 \,|\mathcal{F}_s\right] \\ &\quad + 2\mathbb{E}\left[(W_t - W_s)\, W_s \,|\mathcal{F}_s\right] \\ &= \mathbb{E}\left[(W_t - W_s)^2\right] + W_s^2 + 2W_s\mathbb{E}\left[W_t - W_s\right] = t - s + W_s^2\end{aligned}$$

where we have used the properties of conditional expectations and the defining properties of Brownian motion. Using this result, we arrive at

$$\mathbb{E}\left[W_t^2 - t \,|\mathcal{F}_s\right] = W_s^2 - s$$

and the martingale property of $W_t^2 - t$ is proved.

For the second process we have

$$\begin{aligned}\mathbb{E}\left[\left|e^{uW_t - \frac{1}{2}u^2 t}\right|\right] &= \mathbb{E}\left[e^{uW_t - \frac{1}{2}u^2 t}\right] = \frac{e^{-\frac{1}{2}u^2 t}}{\sqrt{2\pi t}} \int_{-\infty}^{+\infty} e^{ux - \frac{x^2}{2t}}\, dx \\ &= \frac{1}{\sqrt{2\pi t}} \int_{-\infty}^{+\infty} e^{-\frac{(x-ut)^2}{2t}}\, dx = 1\end{aligned}$$

and

$$\mathbb{E}\left[e^{uW_t} \,|\mathcal{F}_s\right] = \mathbb{E}\left[e^{u(W_t - W_s)} e^{uW_s} \,|\mathcal{F}_s\right] = e^{uW_s}\mathbb{E}\left[e^{u(W_t - W_s)}\right] = e^{uW_s} e^{\frac{1}{2}u^2(t-s)}$$

where in the last step we have used the moment generating function of $W_t - W_s$. Using the previous calculation we can write

$$\mathbb{E}\left[e^{uW_t - \frac{1}{2}u^2 t} \,|\mathcal{F}_s\right] = e^{uW_s} e^{\frac{1}{2}u^2(t-s)} e^{-\frac{1}{2}u^2 t} = e^{uW_s - \frac{1}{2}u^2 s}$$

which proves that $e^{uW_t - \frac{1}{2}u^2 t}$ is a martingale. □

Exercise 1.13. Let W_t be a Brownian motion. Prove that $X_t = \exp\left(aW_t + bt\right)$ is a martingale if, and only if, $b = -\frac{a^2}{2}$.

Exercise 1.14. Obtain $\mathbb{E}\left[e^{uW_t}\right]$ for any $u \in \mathbb{R}$.

1.5 Brownian Motion as a Gaussian Process

Brownian motion is a Gaussian process, and this property allows us to state the following characterization, valid under the alternative definition of Brownian motion referred to in Remark 1.6.[3] This result is sometimes used as the definition of Brownian motion.

Theorem 1.15. *A stochastic process is a Brownian motion if, and only if, it is a Gaussian process with zero mean and covariance* $c(s,t) = s \wedge t := \min(s,t)$.

Proof. Let W be a Brownian motion and consider $t_0 < t_1 < \cdots < t_n$ for some $n \in \mathbb{N}$. It is immediate that we can write

$$\begin{pmatrix} W_{t_0} \\ W_{t_1} \\ \vdots \\ W_{t_{n-1}} \\ W_{t_n} \end{pmatrix} = \begin{pmatrix} 1 & 0 & 0 & \cdots & 0 \\ 1 & 1 & 0 & \cdots & 0 \\ \cdots & \cdots & \cdots & \cdots & \cdots \\ 1 & 1 & 1 & 1 & 0 \\ 1 & 1 & 1 & 1 & 1 \end{pmatrix} \begin{pmatrix} W_{t_0} \\ W_{t_1} - W_{t_0} \\ \vdots \\ W_{t_{n-1}} - W_{t_{n-2}} \\ W_{t_n} - W_{t_{n-1}} \end{pmatrix}$$

which is a linear transformation of the vector of Brownian motion increments $(W_{t_0}, W_{t_1} - W_{t_0}, \cdots, W_{t_{n-1}} - W_{t_{n-2}}, W_{t_n} - W_{t_{n-1}})^\top$. As these increments are independent Gaussian random variables, $(W_{t_0}, \ldots, W_{t_n})^\top$ has a multivariate Gaussian distribution, and then W is a Gaussian process. We know from i) of Proposition 1.10 that $\mu(t) = \mathbb{E}[W_t] = 0$. For the covariance, if $s \le t$ we have

$$\begin{aligned} c(s,t) &= \mathbb{E}[(W_s - \mu_s)(W_t - \mu_t)] = \mathbb{E}[W_s W_t] = \mathbb{E}\left[W_s^2 + W_s W_t - W_s^2\right] \\ &= \mathbb{E}\left[W_s^2\right] + \mathbb{E}[W_s(W_t - W_s)] = \mathbb{E}\left[W_s^2\right] + \mathbb{E}[W_s]\,\mathbb{E}[W_t - W_s] \\ &= \mathbb{E}\left[W_s^2\right] = s \end{aligned}$$

where we have used ii) of Proposition 1.10 and that the increment $W_t - W_s$ is independent of $W_s = W_s - W_0$.

Conversely, the increments are normally distributed with zero mean and variance

$$\begin{aligned} \operatorname{Var}(W_t - W_s) &= \operatorname{Cov}(W_t - W_s, W_t - W_s) = c(t,t) - c(t,s) - c(s,t) + c(s,s) \\ &= t - s - s + s = t - s \end{aligned}$$

The covariance of any two non-overlapped increments $W_t - W_s$ and $W_q - W_p$ with $p < q \le s < t$, is given by

$$\operatorname{Cov}\left(W_t - W_s, W_q - W_p\right) = c(t,q) - c(t,p) - c(s,q) + c(s,p) = q - p - q + p = 0$$

As $W_t - W_s$ and $W_q - W_p$ are normally distributed, their zero covariance implies independence. Finally, as discussed in Remark 1.7, the process admits a continuous modification. □

Exercise 1.16. Let W_t be a Brownian motion. Find the covariance function of the process $X_t = W_t^2 - t$.

Exercise 1.17. Let W_t be a Brownian motion and consider, for $\alpha > 0$, the process $X_t = e^{-\alpha t/2} W_{e^{\alpha t}}$. Determine the mean and the covariance function of X_t.

[3]Strictly speaking, in view of Remark 1.6, the following characterization is to be understood up to modifications.

1.6 Transformations of Brownian Motion

There are certain transformations that can be applied to a Brownian motion such that the resulting stochastic process is itself a Brownian motion. These transformations appear in the following result.

Proposition 1.18. *If W_t is a Brownian motion, so are the processes obtained from the following transformations:*

i) Scaling: $X_t = \frac{1}{\sqrt{c}} W_{ct}$, $c > 0$, $t \geq 0$

ii) Time-inversion: $Y_t = \begin{cases} tW_{1/t} & \text{if } t > 0 \\ 0 & \text{if } t = 0 \end{cases}$

iii) Time-reversal: $Z_t = W_T - W_{T-t}$, $0 \leq t \leq T$

iv) Symmetry: $-W_t$, $t \geq 0$

Proof. By Theorem 1.15, we just need to prove that the processes are Gaussian with zero mean and covariance function $c(s,t) = s \wedge t$. In order to prove the Gaussian property, it is enough to show that every finite-dimensional random vector can be obtained as an affine transformation of a Gaussian random vector (see Theorem B.96).
i) Scaling:

$$\begin{pmatrix} X_{t_1} \\ \vdots \\ X_{t_n} \end{pmatrix} = \operatorname{diag}\left(1/\sqrt{c}, \ldots, 1/\sqrt{c}\right) \begin{pmatrix} W_{ct_1} \\ \vdots \\ W_{ct_n} \end{pmatrix}$$

$$\mathbb{E}[X_t] = \frac{1}{\sqrt{c}} \mathbb{E}[W_{ct}] = 0$$

$$\operatorname{Cov}(X_s, X_t) = \frac{1}{c} \operatorname{Cov}(W_{cs}, W_{ct}) = \frac{1}{c} [(cs) \wedge (ct)] = s \wedge t$$

ii) Time-inversion: For $t_1, \ldots, t_n > 0$, we have

$$\begin{pmatrix} Y_{t_1} \\ \vdots \\ Y_{t_n} \end{pmatrix} = \operatorname{diag}(t_1, \ldots, t_n) \begin{pmatrix} W_{1/t_1} \\ \vdots \\ W_{1/t_n} \end{pmatrix}$$

$$\mathbb{E}[Y_t] = t\mathbb{E}\left[W_{1/t}\right] = 0$$

$$\operatorname{Cov}(Y_s, Y_t) = st \operatorname{Cov}\left(W_{1/s}, W_{1/t}\right) = st\left(\frac{1}{s} \wedge \frac{1}{t}\right) = \frac{st}{s \vee t} = s \wedge t$$

For the continuity at $t = 0$, we have that

$$\lim_{t \to 0^+} Y_t = \lim_{t \to 0^+} tW_{1/t} = \lim_{u \to \infty} \frac{W_u}{u} = 0 \text{ a.s.}^4$$

[4] $s \vee t = \max(s, t)$.

where we have used the Strong Law of Large Numbers for Brownian motion.[5]
iii) Time-reversal:

$$\begin{pmatrix} Z_{t_1} \\ Z_{t_2} \\ \vdots \\ Z_{t_{n-1}} \\ Z_{t_n} \end{pmatrix} = \begin{pmatrix} W_T - W_{T-t_1} \\ W_T - W_{T-t_2} \\ \vdots \\ W_T - W_{T-t_{n-1}} \\ W_T - W_{T-t_n} \end{pmatrix} = \begin{pmatrix} 1 & 0 & 0 & \cdots & 0 \\ 1 & 1 & 0 & \cdots & 0 \\ \cdots & \cdots & \cdots & \cdots & \cdots \\ 1 & 1 & 1 & 1 & 0 \\ 1 & 1 & 1 & 1 & 1 \end{pmatrix} \begin{pmatrix} W_T - W_{T-t_1} \\ W_{T-t_1} - W_{T-t_2} \\ \vdots \\ W_{T-t_{n-2}} - W_{T-t_{n-1}} \\ W_{T-t_{n-1}} - W_{T-t_n} \end{pmatrix}$$

$$\mathbb{E}\left[Z_t\right] = \mathbb{E}\left[W_T - W_{T-t}\right] = 0$$

$$\begin{aligned} \mathrm{Cov}\left[Z_s, Z_t\right] &= \mathbb{E}\left[Z_s Z_t\right] = \mathbb{E}\left[\left(W_T - W_{T-s}\right)\left(W_T - W_{T-t}\right)\right] \\ &= \mathbb{E}\left[W_T^2\right] - \mathbb{E}\left[W_T W_{T-t}\right] - \mathbb{E}\left[W_{T-s} W_T\right] + \mathbb{E}\left[W_{T-s} W_{T-t}\right] \\ &= T - T \wedge (T-t) - (T-s) \wedge T + (T-s) \wedge (T-t) \\ &= T - (T-t) - (T-s) + T - (s \vee t) = s + t - s \vee t \\ &= s \wedge t \end{aligned}$$

where we have used that $(T-s) \wedge (T-t) = T - (s \vee t)$ and $s + t - s \vee t = s \wedge t$.
iv) Symmetry:

$$\begin{pmatrix} -W_{t_1} \\ \vdots \\ -W_{t_n} \end{pmatrix} = \mathrm{diag}\,(-1, \ldots, -1) \begin{pmatrix} W_{t_1} \\ \vdots \\ W_{t_n} \end{pmatrix}$$

$$\mathbb{E}\left[-W_t\right] = -\mathbb{E}\left[W_t\right] = 0$$

$$\mathrm{Cov}\left(-W_s, -W_t\right) = \mathrm{Cov}\left(W_s, W_t\right) = s \wedge t$$

□

Exercise 1.19. Let W_t be a Brownian motion. Determine which of the following processes are Brownian motions.

a) $X_t = 3W_{t/9}$

b) $Y_t = W_{3t} - W_t$

c) $Z_t = \sqrt{t} W_1$

1.7 Sample Path Properties

In this section we study the properties of Brownian motion that are related to its sample paths, that is, the mappings $t \mapsto W_t(\omega)$, $\omega \in \Omega$. Brownian motion paths possess many interesting properties, some of them surprising. We analyze in detail those properties that are important for applications. The remaining ones will be stated without proof, together with references for the interested reader.

---

[5]The *Strong Law of Large Numbers* state that $\lim_{t\to\infty} \frac{W_t}{t} = 0$ almost surely (see, for example, Karatzas and Shreve (1991), Problem 9.3 of Chapter 2).

Total and Quadratic Variation

The total amount by which a function increases and decreases over an interval is measured by its total variation, defined as follows.

Definition 1.20. (Total variation, finite variation) The total variation or first-order variation of a function $f : [0,T] \to \mathbb{R}$ is given by

$$TV(f) = \lim_{\|\Pi_n\| \to 0} \sum_{j=0}^{n-1} \left| f(t_{j+1}) - f(t_j) \right| \tag{1.5}$$

where $\Pi_n = \{t_0, t_1, \ldots, t_n\}$ is a partition of $[0,T]$ and $\|\Pi_n\| = \max_{j \in \{0,\ldots,n-1\}} (t_{j+1} - t_j)$. The limit is taken as the length of the longest subinterval goes to 0 (and therefore the number of partition points goes to infinity). We say that f is of finite variation if the limit in (1.5) is finite.

The total variation of a smooth function is finite, and can be computed easily, as the following result shows.

Proposition 1.21. *If $f : [0,T] \to \mathbb{R}$ is differentiable with continuous derivative, its total variation can be obtained as*

$$TV(f) = \int_0^T \left| f'(t) \right| dt$$

Proof. Applying the Mean Value Theorem to the interval $[t_j, t_{j+1}]$ of the partition Π_n, we have that there is a $t_j^* \in (t_j, t_{j+1})$ such that

$$\frac{f(t_{j+1}) - f(t_j)}{t_{j+1} - t_j} = f'\left(t_j^*\right)$$

Replacing this equality in expression (1.5), we obtain

$$TV(f) = \lim_{\|\Pi_n\| \to 0} \sum_{j=0}^{n-1} \left| f'\left(t_j^*\right) \right| (t_{j+1} - t_j) = \int_0^T \left| f'(t) \right| dt$$

where we have applied the definition of the Riemann integral. □

Unlike what happens with C^1-functions, the total variation of Brownian motion paths is not finite (see Theorem 1.29 below). We also have the following result.

Proposition 1.22. *If $f : [0,T] \to \mathbb{R}$ is monotone, then it is of finite variation.*

Proof. If f is increasing, it satisfies $f(s) \leq f(t)$, $s < t$. Then, considering a partition $\{t_0, \ldots, t_n\}$ of $[0,T]$, we have

$$\sum_{j=0}^{n-1} \left| f(t_{j+1}) - f(t_j) \right| = \sum_{j=0}^{n-1} \left(f(t_{j+1}) - f(t_j) \right) = f(t_n) - f(t_0)$$

which is independent of the partition. Thus, we have $TV(f) = f(t_n) - f(t_0)$, which is finite. The proof for the case in which f is decreasing is analogous. □

A more appropriate concept for stochastic processes than that of total variation is the one defined below.

Definition 1.23. (Quadratic variation) The quadratic variation of a function $f : [0,T] \to \mathbb{R}$ is defined as

$$[f,f](T) = \lim_{\|\Pi_n\| \to 0} \sum_{j=0}^{n-1} \left[f(t_{j+1}) - f(t_j) \right]^2$$

Proposition 1.24. *The quadratic variation of a continuously differentiable function is zero.*

Proof. Using the Mean Value Theorem as in the proof of Proposition 1.21, we can write

$$\begin{aligned}
[f,f](T) &= \lim_{\|\Pi_n\| \to 0} \sum_{j=0}^{n-1} \left| f'\left(t_j^*\right) \right|^2 (t_{j+1} - t_j)^2 \\
&\leq \lim_{\|\Pi_n\| \to 0} \left[\|\Pi_n\| \sum_{j=0}^{n-1} \left| f'\left(t_j^*\right) \right|^2 (t_{j+1} - t_j) \right] \\
&= \left(\lim_{\|\Pi_n\| \to 0} \|\Pi_n\| \right) \int_0^T \left| f'(t) \right|^2 dt = 0
\end{aligned}$$

where in the last step we have used that $\int_0^T |f'(t)|^2 \, dt$ is finite due to the continuity of f'. □

The fact that differentiable functions with continuous derivative have zero quadratic variation is the reason why this concept rarely appears in calculus textbooks. However, as we will see below, the quadratic variation of Brownian motion is not zero, and this result is crucial to the emergence of stochastic calculus.

Theorem 1.25. *If W is a Brownian motion, then $[W,W](T) = T$ in L^2 sense.*

Proof. Define $Q_{\Pi_n} = \sum_{j=0}^{n-1} \left(W_{t_{j+1}} - W_{t_j} \right)^2$. We need to prove that

$$\lim_{\|\Pi_n\| \to 0} \mathbb{E}\left[\left(Q_{\Pi_n} - T \right)^2 \right] = 0$$

By the definition of Q_{Π_n} we have

$$Q_{\Pi_n} - T = \sum_{j=0}^{n-1} \left(W_{t_{j+1}} - W_{t_j} \right)^2 - T = \sum_{j=0}^{n-1} \left[\left(W_{t_{j+1}} - W_{t_j} \right)^2 - (t_{j+1} - t_j) \right] := \sum_{j=0}^{n-1} Y_j$$

where Y_j are independent random variables with $\mathbb{E}[Y_j] = 0$, $j = 0, \dots, n-1$. Therefore

$$\begin{aligned}
\mathbb{E}\left[(Q_{\Pi_n} - T)^2\right] &= \mathbb{E}\left[\left(\sum_{j=0}^{n-1} Y_j\right)^2\right] = \operatorname{var}\left(\sum_{j=0}^{n-1} Y_j\right) = \sum_{j=0}^{n-1} \operatorname{var}(Y_j) = \sum_{j=0}^{n-1} \mathbb{E}\left(Y_j^2\right) \\
&= \sum_{j=0}^{n-1} \mathbb{E}\left[\left(W_{t_{j+1}} - W_{t_j}\right)^4\right] - 2\sum_{j=0}^{n-1} (t_{j+1} - t_j)\, \mathbb{E}\left[\left(W_{t_{j+1}} - W_{t_j}\right)^2\right] \\
&\quad + \sum_{j=0}^{n-1} (t_{j+1} - t_j)^2 \\
&= 3\sum_{j=0}^{n-1} (t_{j+1} - t_j)^2 - 2\sum_{j=0}^{n-1} (t_{j+1} - t_j)^2 + \sum_{j=0}^{n-1} (t_{j+1} - t_j)^2 \\
&= 2\sum_{j=0}^{n-1} (t_{j+1} - t_j)^2 \le 2\,\|\Pi_n\| \sum_{j=0}^{n-1} (t_{j+1} - t_j) = 2\,\|\Pi_n\|\, T
\end{aligned}$$

Taking limits we finally arrive at

$$\lim_{\|\Pi_n\| \to 0} \mathbb{E}\left[(Q_{\Pi_n} - T)^2\right] = 0$$

□

Remark 1.26. L^2-convergence in Theorem 1.25 guarantees convergence in probability. If we add some additional hypotheses on the partition, a.s. convergence is also achieved (see Protter (2005), page 18).

Remark 1.27. Theorem 1.25 states that Brownian motion accumulates T units of quadratic variation on $[0, T]$ or, according to Shreve (2004): "Brownian motion accumulates quadratic variation at rate one per unit time".

Before proving that Brownian motion has infinite total variation, we need the following result, that relates the concepts of total and quadratic variation for continuous functions.

Theorem 1.28. *If $f : [0, T] \to \mathbb{R}$ is continuous and of finite variation, then its quadratic variation is zero.*

Proof. From Definition 1.23 we have

$$\begin{aligned}
[f, f](T) &= \lim_{\|\Pi_n\| \to 0} \sum_{j=0}^{n-1} \left[f(t_{j+1}) - f(t_j)\right]^2 \\
&\le \lim_{\|\Pi_n\| \to 0} \max_j \left|f(t_{j+1}) - f(t_j)\right| \sum_{j=0}^{n-1} \left|f(t_{j+1}) - f(t_j)\right| \\
&= TV(f) \lim_{\|\Pi_n\| \to 0} \max_j \left|f(t_{j+1}) - f(t_j)\right| = 0
\end{aligned}$$

where in the last step we have applied that f is uniformly continuous, as a continuous function on a compact interval. □

Theorem 1.29. *Almost every Brownian motion path has infinite variation.*

Proof. By iv) of Definition 1.5, Brownian motion paths are a.s. continuous. As by Theorem 1.25 the quadratic variation of Brownian motion is not zero (see also Remark 1.26), by Theorem 1.28 we conclude that Brownian motion paths have infinite variation a.s. □

Considering the interpretation of the total variation as the accumulated displacement of the Brownian motion path in the interval, the above result can be stated by saying that the Brownian motion travels an infinite length in any time interval, however small it may be.

Another striking result about Brownian motion is the following.

Proposition 1.30. *Almost every Brownian motion path is not monotone in any interval.*

Proof. Immediate from Proposition 1.22 and Theorem 1.29. □

Further Sample Path Properties

Although we have already studied the properties of Brownian motion paths that are most relevant for our purposes, we do not want to conclude this section without mentioning other striking properties. References for all the properties listed below can be found in Karatzas and Shreve (1991), Chapter 2.

- With probability one, a Brownian motion path changes sign infinitely many times on every time interval $[0, \varepsilon]$, $\varepsilon > 0$ (Problem 7.18, with solution).
- The probability that Brownian motion returns to the origin infinitely often is one (Problem 9.5, with solution).
- For almost every Brownian path, the set of points of local maximum is countable and dense in $[0, \infty)$, and all local maxima are strict (Theorem 9.12).
- Almost every Brownian path has no point of increase (or decrease) (Theorem 9.13).
- Almost every sample path of Brownian motion is nowhere differentiable (Theorem 9.18).

Fractal Nature of Brownian Motion*

Some of the properties of Brownian motion paths that we have studied, such as the scaling property (i) of Proposition 1.18, their infinite total variation over finite time intervals (Theorem 1.29) or their highly irregular behavior (see the previous subsection), are characteristic of the sets known as fractals. In fact, Mandelbrot himself, known as the father of fractal geometry, presents Brownian motion (both mathematical and natural) as an example of a fractal: "For example, Brownian curves are fractal sets, and physical Brownian motion is a natural fractal" (Mandelbrot (1982), page 5).

To establish more rigorously the fractal nature of Brownian paths, we need a formal definition of fractal (Haussermann, 1989).

Definition 1.31. (Fractal) A subset, E, of $\mathbb{R}^n$ is a fractal if the topological dimension of E is different from the Hausdorff dimension of E.[6]

The graph of a Brownian motion path has topological dimension 1 almost surely, as is the case for the graph of any continuous function. On the other hand, it is possible to prove that the Hausdorff dimension of the graph of a Brownian motion path is $3/2$ with probability 1 (Falconer, 2003). Considering these two facts and Definition 1.31, we can state the following result.

Theorem 1.32. *The graph of almost every Brownian motion path is a fractal set with Hausdorff dimension 3/2.*

1.8 Markov and Strong Markov Properties

Intuitively, a process is said to satisfy the *Markov property*, or to be a *Markov process*, if the future behavior of the process, given its current state, is independent of the past. In other words, it is as if the process "does not remember" its past. Hence, Markov processes are sometimes described as "memoryless" processes.

Taking into account our interpretation of filtration, this intuitive idea of Markov processes can be formalized as follows.

Definition 1.33. (Markov process) The adapted process X is a Markov process if for every $s, t > 0$ and every $y \in \mathbb{R}$ it satisfies

$$\mathbb{P}\left[X_{t+s} \leq y \,|\mathcal{F}_t\right] = \mathbb{P}\left[X_{t+s} \leq y \,|X_t\right] \quad \text{a.s.} \tag{1.6}$$

i.e., the distributions of future values of the process conditional on past information and to its current value are a.s. the same.

Theorem 1.34. *Brownian motion is a Markov process.*

Proof. Since moment generating functions uniquely determine probability distributions, it is enough to prove that $\mathbb{E}\left[\exp\left(uW_{t+s}\right)|\mathcal{F}_t\right] = \mathbb{E}\left[\exp\left(uW_{t+s}\right)|W_t\right]$ for $s, t > 0$, where W is a Brownian motion. Applying the independence from the past of Brownian motion increments, we obtain the following chain of equalities that conclude the proof

$$\begin{aligned}
\mathbb{E}\left[e^{uW_{t+s}}\,|\mathcal{F}_t\right] &= \mathbb{E}\left[e^{u(W_{t+s}-W_t+W_t)}\,|\mathcal{F}_t\right] = e^{uW_t}\mathbb{E}\left[e^{u(W_{t+s}-W_t)}\,|\mathcal{F}_t\right] \\
&= e^{uW_t}\mathbb{E}\left[e^{u(W_{t+s}-W_t)}\right] = e^{uW_t}\mathbb{E}\left[e^{u(W_{t+s}-W_t)}\,|W_t\right] \\
&= \mathbb{E}\left[e^{uW_{t+s}}\,|W_t\right]
\end{aligned}$$

□

As we have just seen, Brownian motion possesses the Markov property (1.6) for any deterministic times $s, t > 0$. Another characteristic that makes Brownian motion special is that this property can be generalized to a special type of random times, called *stopping times*, giving rise to the *strong Markov property*. Before stating it, we need to formalize some of these concepts.

[6]The concepts of topological dimension and Hausdorff dimension are beyond the reach of this book. We refer the reader to Haussermann (1989) for the corresponding definitions.

Definition 1.35. (Random time) A random time is a random variable taking values in $[0, \infty]$.

In practice, random times are assigned to instants determined by the outcome of a random experiment, such as the moment it starts to rain or the moment when a stock price exceeds a certain value.

Definition 1.36. (Stopping time) A random time T is a stopping time if the event $\{T \leq t\}$ belongs to $\mathcal{F}_t$ for every $t \geq 0$.

Intuitively, a random time T is a stopping time if at each instant t it is possible to decide whether T has occurred or not, with the information available up to that instant. As an immediate consequence of Definition 1.36 we obtain that every random time equal to a nonnegative constant is a stopping time.

The previous examples of random times are obviously stopping times. An example of a random time T that is not a stopping time is the last zero of a Brownian motion before $t = 1$. It is not a stopping time since the information needed to decide whether T has occurred or not at time $t < 1$ is contained in $\mathcal{F}_1$, but not in $\mathcal{F}_t \subset \mathcal{F}_1$ (we do not know if a zero is the last one before $t = 1$ until that instant of time).

The following result states that the process formed by the increments of a Brownian motion from a stopping time is itself a Brownian motion.

Theorem 1.37. *Let W be a Brownian motion and T an a.s. finite stopping time. Then, the process $W_t^{(T)} := W_{T+t} - W_T$, $t \geq 0$ is a Brownian motion independent of $\mathcal{F}_T$.*

Proof. See Rogers and Williams (1994), Theorem 12.1, page 21. □

Remark 1.38. In Theorem 1.37, the condition that T is a stopping time cannot be relaxed, i.e., if we only consider random times, the theorem ceases to be valid.

Theorem 1.37 is usually explained in intuitive terms by saying that Brownian motion "starts afresh" at stopping times, i.e., the process $W^{(T)}$ is a Brownian motion started at zero.

The strong Markov property can be considered as a generalization of the Markov property to stopping times.

Definition 1.39. (Strong Markov process) The adapted process X is a strong Markov process if it satisfies

$$\mathbb{P}\left[X_{T+t} \leq y \,|\, \mathcal{F}_T\right] = \mathbb{P}\left[X_{T+t} \leq y \,|\, X_T\right] \quad \text{a.s.}$$

for every $t \geq 0$, $y \in \mathbb{R}$ and every finite stopping time T.

Brownian motion satisfies this property, as stated in the following corollary.

Corollary 1.40. ***(Strong Markov property of Brownian motion)*** *Brownian motion is a strong Markov process.*[7]

Proof. It suffices to follow the same steps of the proof of Theorem 1.34 and apply Theorem 1.37. □

[7] In some texts, Theorem 1.37 is taken as the definition of the strong Markov property of Brownian motion.

1.9 Hitting Times and the Reflection Principle

In financial applications of stochastic calculus, it is sometimes interesting to know the distribution of the first random time at which the price of an asset reaches a certain level (for example, in the context of barrier option valuation). We will study these times, called *hitting times*, or *first passage times*, for the case of Brownian motion.

Definition 1.41. (Hitting time) The hitting time, T_b, of a Brownian motion W to a level $b \in \mathbb{R}$ is defined as

$$T_b(\omega) = \inf\{t \geq 0 : W_t(\omega) = b\}$$

For a general stochastic process, the hitting time of a given level does not have to be a stopping time. However, for Brownian motion, the continuity of its sample paths allows us to state the following result.

Proposition 1.42. *The hitting times of a Brownian motion are stopping times.*

Proof. See Karatzas and Shreve (1991), Problem 2.7, page 7. □

The following theorem presents the *Reflection Principle*, an important consequence of the strong Markov property that plays a key role in the study of Brownian motion.

Theorem 1.43. ***(Reflection Principle)*** *If W is a Brownian motion and T is an a.s. finite stopping time, then the process $\overline{W}^T$ defined by*

$$\overline{W}_t^T = \begin{cases} W_t & \text{if } t \leq T \\ 2W_T - W_t & \text{if } t > T \end{cases} \tag{1.7}$$

is also a Brownian motion. The process $\overline{W}^T$ is called Brownian motion reflected at T.

Proof. See Schilling and Partzsch (2012), Theorem 6.12. □

The name "Brownian motion reflected at T" comes from the fact that the sample paths of $\overline{W}_t^T$ coincide with those of W up to the stopping time T, and from that time on, they are the "reflection" of the paths of W at the level W_T (see Figure 1.2).

One of the most interesting consequences of the Reflection Principle is an expression for the density function of hitting times for Brownian motion. Before stating this result, we need to introduce another important process related to Brownian motion.

1.10 The Running Maximum of Brownian Motion

Definition 1.44. (Running maximum of Brownian motion) Given a Brownian motion W, its running maximum, M, is the stochastic process whose value at time t is the maximum value of W up to time t

$$M_t = \max_{0 \leq s \leq t} W_s$$

Theorem 1.45. *For any $b > 0$*

$$\mathbb{P}[M_t > b] = 2\mathbb{P}[W_t > b] = 2\left[1 - \Phi\left(\frac{b}{\sqrt{t}}\right)\right]$$

where Φ is the standard Gaussian distribution function.

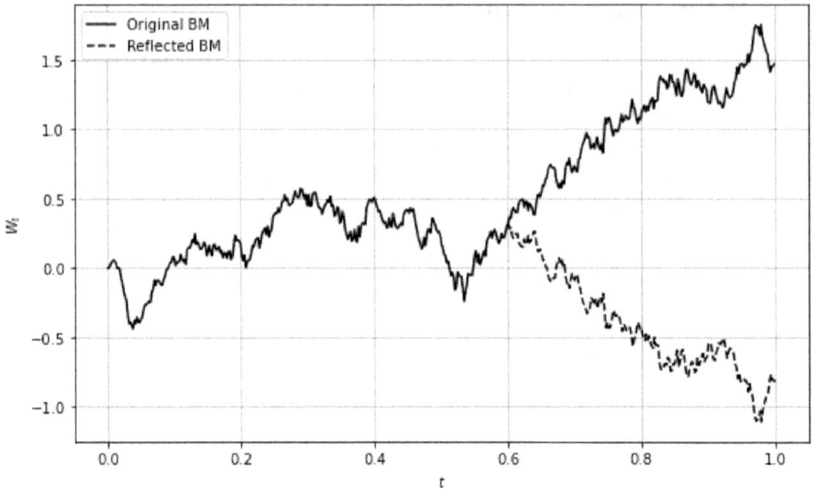

Figure 1.2: A Brownian motion and its reflected path.

Proof. Let T_b be the hitting time of the Brownian motion W to the level b. We know from Proposition 1.42 that T_b is a stopping time. To prove the first equality note that

$$\begin{aligned}\mathbb{P}[M_t > b] &= \mathbb{P}[\{M_t > b, W_t > b\} \cup \{M_t > b, W_t < b\}] \\ &= \mathbb{P}[\{W_t > b\} \cup \{M_t > b, W_t < b\}] \\ &= \mathbb{P}\left[\{W_t > b\} \cup \left\{\overline{W}_t^{T_b} > b\right\}\right] \\ &= 2\mathbb{P}[W_t > b]\end{aligned}$$

where we have applied that $\{W_t > b\} \subset \{M_t > b\}$, the definition of $\overline{W}_t^T$ (expression (1.7)) and the Reflection Principle (Theorem 1.43).

The second equality comes from the following calculation

$$\mathbb{P}[W_t > b] = \frac{1}{\sqrt{2\pi t}} \int_b^\infty e^{-x^2/2t} dx = \frac{1}{\sqrt{2\pi}} \int_{b/\sqrt{t}}^\infty e^{-u^2/2} du$$

where we have used the density function of a Brownian motion (expression (1.4)). □

Remark 1.46. In the proof of Theorem 1.45 we have utilized that, as a consequence of the Reflection Principle, we get

$$\mathbb{P}[M_t > b, W_t < b] = \mathbb{P}\left[\overline{W}_t^{T_b} > b\right] = \mathbb{P}[W_t > b]$$

This fact can be intuitively interpreted as if for any Brownian motion path that starts from a hitting time, there is another path, obtained by reflection at level b, and these two paths have the same “probability”.[8] This intuition is usually taken as the heuristic rationale behind the Reflection Principle (see Karatzas and Shreve (1991)).

Now we can state the result with the density function of hitting times.

Proposition 1.47. *The density function of T_b is given by*

$$\rho_b(t) = \frac{|b|}{\sqrt{2\pi t^3}} e^{-b^2/2t}$$

which is the density function of an Inverse Gamma distribution with shape parameter $\frac{1}{2}$ and scale parameter $\frac{b^2}{2}$.[9]

Proof. Taking into account that $\{T_b < t\} = \{M_t > b\}$, from Theorem 1.45 we have that for $b > 0$

$$\mathbb{P}[T_b < t] = \mathbb{P}[M_t > b] = \frac{2}{\sqrt{2\pi}} \int_{b/\sqrt{t}}^{\infty} e^{-u^2/2} du \tag{1.8}$$

Differentiating this expression with respect to t we obtain the desired result. For the case $b < 0$, it suffices to apply Theorem 1.45 to the Brownian motion $-W$ and the level $-b > 0$. □

As a consequence of Proposition 1.47, we have the following result, which has a counterintuitive interpretation: Brownian motion reaches any level b with probability one, but the expected time for this event to happen is infinite.

Corollary 1.48. *The hitting time T_b of a Brownian motion satisfies $\mathbb{P}[T_b < \infty] = 1$ and $\mathbb{E}[T_b] = \infty$.*

Proof. From expression (1.8) we have

$$\mathbb{P}[T_b < \infty] = \frac{2}{\sqrt{2\pi}} \int_0^{\infty} e^{-u^2/2} du = 1$$

For the second equality

$$\mathbb{E}[T_b] = \int_0^{\infty} \frac{|b|\, t}{\sqrt{2\pi t^3}} e^{-b^2/2t} dt = \frac{|b|}{\sqrt{2\pi}} \int_0^{\infty} \frac{1}{\sqrt{t}} e^{-b^2/2t} dt$$

To prove that this integral diverges, since $\lim_{t\to\infty} e^{-b^2/2t} = 1$, we can find $T > 0$ such that $e^{-b^2/2t} \geq 1/2$. Therefore, we have

$$\mathbb{E}[T_b] \geq \frac{|b|}{\sqrt{2\pi}} \int_T^{\infty} \frac{1}{2\sqrt{t}} dt = \frac{|b|}{\sqrt{2\pi}} \left[\sqrt{t}\right]_T^{\infty} = \infty$$

□

[8] Of course, this explanation only makes sense as a heuristic argument, since the probability that a given sample path occurs is zero.

[9] The Inverse Gamma distribution has a density function given by $f(t; \alpha, \beta) = \frac{\beta^\alpha}{\Gamma(\alpha)} \frac{1}{t^{\alpha+1}} \exp\left(-\frac{\beta}{t}\right)$ where α is the shape parameter, β is the scale parameter, and Γ is the Gamma function.

As T_b has infinite expectation, it is sometimes useful to work with its Laplace transform. We have the following result.

Proposition 1.49. *For* $\alpha, b > 0$

$$\mathbb{E}\left[e^{-\alpha T_b}\right] = e^{-b\sqrt{2\alpha}}$$

Proof.

$$\mathbb{E}\left[e^{-\alpha T_b}\right] = \frac{|b|}{\sqrt{2\pi}} \int_0^\infty \frac{1}{\sqrt{t^3}} e^{-\alpha t - b^2/2t} dt = \frac{2}{\sqrt{2\pi}} \int_0^\infty e^{-\left(\alpha b^2/u^2 + u^2/2\right)} du = e^{-b\sqrt{2\alpha}}$$

where in the last step we have used the equality

$$\int_0^\infty e^{-\left(ax^2 + b/x^2\right)} dx = \frac{1}{2}\sqrt{\frac{\pi}{a}} e^{-2\sqrt{ab}}$$

□

For some applications (e.g., barrier option valuation), it is useful to know the joint distribution of a Brownian motion and its running maximum. The following result provides the density function of this distribution.

Proposition 1.50. *For* $t > 0$ *and* $a \leq b$, $b \geq 0$

$$\mathbb{P}\left[W_t \in da, M_t \in db\right] = \frac{2\left(2b - a\right)}{\sqrt{2\pi t^3}} \exp\left\{-\frac{(2b-a)^2}{2t}\right\} dadb$$

Proof. Using that $\{M_t > b\} = \{T_b < t\}$ we can write

$$\begin{aligned}
\mathbb{P}\left[W_t < a, M_t > b\right] &= \mathbb{P}\left[\overline{W}_t^{T_b} > 2b - a, M_t > b\right] \\
&= \mathbb{P}\left[W_t > 2b - a, M_t > b\right] \\
&= \mathbb{P}\left[W_t > 2b - a\right] \\
&= \frac{1}{\sqrt{2\pi t}} \int_{2b-a}^\infty e^{-x^2/2t} dx
\end{aligned}$$

where we have used the Reflection Principle and that $\{W_t > 2b - a\} \subset \{M_t > b\}$ since $2b - a \geq b$. Taking the partial derivative $\frac{\partial^2}{\partial a \partial b}$ we obtain the desired result. □

1.11 Brownian Motion in Higher Dimensions

Some of the most important models in quantitative finance work with multiple sources of randomness associated with different Brownian motions, both independent and correlated. In this section, we present the mathematics behind these models.

The following two definitions are natural extensions of Definitions 1.1 and 1.5 to the multidimensional framework.

Definition 1.51. (Multidimensional stochastic process) A (continuous-time) d-dimensional stochastic process $\mathbf{X}$ is a collection $\mathbf{X} = \{\mathbf{X}_t\}_{t\geq 0}$ of random d-vectors on $(\Omega, \mathcal{F})$.

Definition 1.52. (Multidimensional Brownian motion) A d-dimensional stochastic process $\mathbf{W}$ is a d-dimensional Brownian motion if it satisfies:

i) (Adapted) $\mathbf{W}_t$ is a $\mathcal{F}_t$-measurable random vector for each $t \geq 0$.

ii) (Increments independent of the past) For $0 \leq s < t$ the increment $\mathbf{W}_t - \mathbf{W}_s$ is independent of $\mathcal{F}_s$.

iii) (Gaussian increments) For $0 \leq s < t$ the increment $\mathbf{W}_t - \mathbf{W}_s$ is normally distributed with $\mathbf{W}_t - \mathbf{W}_s \sim N(\mathbf{0}, (t-s)\mathbf{I})$, where $\mathbf{0} \in \mathbb{R}^d$ and $\mathbf{I}$ is the identity matrix of order d.

iv) (Continuous paths) The function $t \mapsto \mathbf{W}_t(\omega)$ is continuous for almost every $\omega \in \Omega$.

Remark 1.53. Definition 1.52 is coherent with Definition 1.5, namely, when $d = 1$, both definitions coincide.

In what follows, we will state that a d-dimensional stochastic process is a d-dimensional Brownian motion if, and only if, its coordinate processes are independent one-dimensional Brownian motions. To prove this result we need the following characterization of d-dimensional Brownian motion.

Lemma 1.54. *A d-dimensional stochastic process $\mathbf{X}$ with a.s. continuous sample paths is a d-dimensional Brownian motion with respect to $\mathcal{F}_t^{\mathbf{X}}$ if, and only if, it satisfies*

$$\mathbb{E}\left[\exp\left(i\sum_{j=1}^{n}\langle\xi_j, \mathbf{X}_{t_j} - \mathbf{X}_{t_{j-1}}\rangle\right)\right] = \exp\left[-\frac{1}{2}\sum_{j=1}^{n}\|\xi_j\|^2(t_j - t_{j-1})\right] \tag{1.9}$$

for all $n > 0$, $0 = t_0 < t_1 < \cdots < t_n$, and $\xi_1, \ldots, \xi_n \in \mathbb{R}^d$, where $\langle\cdot,\cdot\rangle$ and $\|\cdot\|$ denote, respectively, the inner product and the Euclidean norm in $\mathbb{R}^d$.

Proof. Assume that $\mathbf{X}$ is a d-dimensional Brownian motion. Then we have

$$\begin{aligned}
\mathbb{E}\left[\exp\left(i\sum_{j=1}^{n}\langle\xi_j, \mathbf{X}_{t_j} - \mathbf{X}_{t_{j-1}}\rangle\right)\right] &= \prod_{j=1}^{n}\mathbb{E}\left[\exp\left(i\langle\xi_j, \mathbf{X}_{t_j} - \mathbf{X}_{t_{j-1}}\rangle\right)\right] \\
&= \prod_{j=1}^{n}\exp\left[-\frac{1}{2}\|\xi_j\|^2(t_j - t_{j-1})\right] \\
&= \exp\left[-\frac{1}{2}\sum_{j=1}^{n}\|\xi_j\|^2(t_j - t_{j-1})\right]
\end{aligned}$$

where we have applied the independence of the increments, and that $\mathbf{X}_{t_j} - \mathbf{X}_{t_{j-1}} \sim N(\mathbf{0}, (t_j - t_{j-1})\mathbf{I})$, and therefore, its characteristic function is

$$\begin{aligned}
\mathbb{E}\left[\exp\left(i\langle\xi_j, \mathbf{X}_{t_j} - \mathbf{X}_{t_{j-1}}\rangle\right)\right] &= \exp[-\frac{1}{2}\xi^\top(t_j - t_{j-1})\mathbf{I}\xi] \\
&= \exp\left[-\frac{1}{2}\|\xi_j\|^2(t_j - t_{j-1})\right]
\end{aligned} \tag{1.10}$$

Conversely, assume that (1.9) holds. Condition i) of Definition 1.52 is satisfied by the assumption about the filtration. Fix $k \in \{1, \ldots, n\}$ and let $\xi_j = \mathbf{0}$ for all $j \neq k$. Then, by (1.9) we have

$$\mathbb{E}\left[\exp\left(i\left\langle \xi_k, \mathbf{X}_{t_k} - \mathbf{X}_{t_{k-1}}\right\rangle\right)\right] = \exp\left[-\frac{1}{2}\|\xi_k\|^2(t_k - t_{k-1})\right]$$

which proves iii) of Definition 1.52. On the other hand, expression (1.9) shows that the joint characteristic function factorizes, and therefore, by vi) of Proposition B.92, the increments are independent and hence independent of the past (Karatzas and Shreve (1991), Problem 1.4, page 49). □

Theorem 1.55. *The d-dimensional stochastic process $\mathbf{W} = \left(W^{(1)}, \ldots, W^{(d)}\right)^\top$ is a d-dimensional Brownian motion if, and only if, the processes $W^{(k)}$, $k = 1, \ldots, d$, are independent one-dimensional Brownian motions.*

Proof. Assume $\mathbf{W}$ is a d-dimensional Brownian motion. The continuity of the processes $W^{(k)}$ is guaranteed by the continuity of $\mathbf{W}$. Taking $\xi_j = z_j \mathbf{e}_k \; \forall j \in \{1, \ldots, n\}$, with $z_j \in \mathbb{R}$ and $\mathbf{e}_k$ the k-th element of the canonical basis of $\mathbb{R}^d$, we have by (1.9) that

$$\mathbb{E}\left[\exp\left(i\sum_{j=1}^{n} z_j\left(W_{t_j}^{(k)} - W_{t_{j-1}}^{(k)}\right)\right)\right] = \exp\left[-\frac{1}{2}\sum_{j=1}^{n} z_j^2(t_j - t_{j-1})\right]$$

and then, $W^{(k)}$ is a one-dimensional Brownian motion for $k = 1, \ldots, d$.

On the other hand, using (1.9) with $n = 1$ and $t_1 = t$, and the characteristic function of Brownian motion, we can write

$$\begin{aligned}
\mathbb{E}\left[\exp\left(i\sum_{k=1}^{d} \xi^{(k)} W_t^{(k)}\right)\right] &= \mathbb{E}\left[\exp\left(i\langle \xi, \mathbf{W}_t\rangle\right)\right] = \exp\left[-\frac{1}{2}\|\xi\|^2 t\right] \\
&= \exp\left[-\frac{1}{2}t\sum_{k=1}^{d}\left(\xi^{(k)}\right)^2\right] = \prod_{k=1}^{d}\exp\left[-\frac{1}{2}t\left(\xi^{(k)}\right)^2\right] \\
&= \prod_{k=1}^{d}\mathbb{E}\left[\exp\left(i\xi^{(k)} W_t^{(k)}\right)\right]
\end{aligned}$$

from where we conclude the independence of $W_t^{(k)}$ for $k = 1, \ldots, d$.

Conversely, if $W^{(k)}$, $k = 1, \ldots, d$ are independent, one-dimensional Brownian motions, we have

$$
\begin{aligned}
\mathbb{E}\left[\exp\left(i\sum_{j=1}^{n}\langle\xi_j,\mathbf{W}_{t_j}-\mathbf{W}_{t_{j-1}}\rangle\right)\right] &= \mathbb{E}\left[\exp\left(i\sum_{j=1}^{n}\sum_{k=1}^{d}\xi_j^{(k)}\left(W_{t_j}^{(k)}-W_{t_{j-1}}^{(k)}\right)\right)\right] \\
&= \mathbb{E}\left[\prod_{k=1}^{d}\exp\left(i\sum_{j=1}^{n}\xi_j^{(k)}\left(W_{t_j}^{(k)}-W_{t_{j-1}}^{(k)}\right)\right)\right] \\
&= \prod_{k=1}^{d}\mathbb{E}\left[\exp\left(i\sum_{j=1}^{n}\xi_j^{(k)}\left(W_{t_j}^{(k)}-W_{t_{j-1}}^{(k)}\right)\right)\right] \\
&= \prod_{k=1}^{d}\exp\left[-\frac{1}{2}\sum_{j=1}^{n}\left(\xi_j^{(k)}\right)^2\left(t_j-t_{j-1}\right)\right] \\
&= \exp\left[-\frac{1}{2}\sum_{k=1}^{d}\sum_{j=1}^{n}\left(\xi_j^{(k)}\right)^2\left(t_j-t_{j-1}\right)\right] \\
&= \exp\left[-\frac{1}{2}\sum_{j=1}^{n}\left\|\xi_j\right\|^2\left(t_j-t_{j-1}\right)\right] \qquad (1.11)
\end{aligned}
$$

where we have used that $W^{(1)},\cdots,W^{(d)}$, are independent, and expression (1.10). Since $W^{(k)}$, $k=1,\ldots,d$ are continuous processes, so is $\mathbf{W}=\left(W^{(1)},\ldots,W^{(d)}\right)^{\top}$. Then, by expression (1.11) and Lemma 1.54, $\mathbf{W}$ is a d-dimensional Brownian motion. □

Exercise 1.56. Let $\left(W_t,\widetilde{W}_t\right)$ be a two-dimensional Brownian motion, and let ρ be a real number with $|\rho|\leq 1$.

a) Prove that the process $X_t=\rho W_t+\sqrt{1-\rho^2}\widetilde{W}_t$ is a Brownian motion.

b) Find $\operatorname{Corr}(W_t,X_t)$

In some financial applications of Brownian motion in higher dimensions, it is too restrictive to require one-dimensional Brownian motions to be independent, and it is usual to work with correlated Brownian motion. The mathematical definition of this process is as follows.

Definition 1.57. (Correlated Brownian motion) A d-dimensional process ${}^{\mathbf{Q}}\mathbf{W}=\left({}^{\mathbf{Q}}W^{(1)},\ldots,{}^{\mathbf{Q}}W^{(d)}\right)$, is a correlated Brownian motion with correlation matrix $\mathbf{Q}$, if it satisfies i), ii) and iv) of Definition 1.52 and

iii)′ For $0\leq s<t$, the increment ${}^{\mathbf{Q}}\mathbf{W}_t-{}^{\mathbf{Q}}\mathbf{W}_s$ is normally distributed with

$$
{}^{\mathbf{Q}}\mathbf{W}_t-{}^{\mathbf{Q}}\mathbf{W}_s\sim N\left(\mathbf{0},(t-s)\,\mathbf{Q}\right)
$$

where $\mathbf{0} = (0, \ldots, 0)^{\top}_{1\times d}$ and $\mathbf{Q}$ is the correlation matrix associated to the random vector ${}^{\mathbf{Q}}\mathbf{W}_t - {}^{\mathbf{Q}}\mathbf{W}_s$.

From the definition it is clear that d-dimensional Brownian motion can be considered as a correlated Brownian motion with $\mathbf{Q} = \mathbf{I}$. We will also assume that ${}^{\mathbf{Q}}\mathbf{W}_0 = \mathbf{0}$ a.s.

The following result shows that we can obtain a correlated Brownian motion by means of a linear transformation of a multidimensional Brownian motion.

Proposition 1.58. *Let* $\mathbf{W} = (W^{(1)}, \ldots, W^{(d)})^{\top}$ *be a* d*-dimensional Brownian motion, and* $\boldsymbol{\Sigma}$ *a* $d \times d$ *matrix with elements* $(\boldsymbol{\Sigma})_{ij} = \sigma_{ij}$ *satisfying* $\sum_{j=1}^{d} \sigma_{ij}^2 = 1$ *for* $i = 1, \ldots, d$. *Then,* $\mathbf{X} := \boldsymbol{\Sigma}\mathbf{W}$ *is a correlated Brownian motion with correlation matrix* $\boldsymbol{\Sigma}\boldsymbol{\Sigma}^{\top}$.

Proof. The conditions i), ii) and iv) of Definition 1.57 are automatically satisfied when we replace $\mathbf{W}$ by $\boldsymbol{\Sigma}\mathbf{W}$. For condition iii)' we have

$$\begin{aligned}
\mathbb{E}\left[\exp\left(i\langle \xi, \mathbf{X}_t - \mathbf{X}_s\rangle\right)\right] &= \mathbb{E}\left[\exp\left(i\langle \xi, \boldsymbol{\Sigma}\left(\mathbf{W}_t - \mathbf{W}_s\right)\rangle\right)\right] \\
&= \mathbb{E}\left[\exp\left(i\left\langle \boldsymbol{\Sigma}^{\top}\xi, \mathbf{W}_t - \mathbf{W}_s\right\rangle\right)\right] \\
&= \exp\left[-\frac{1}{2}\left\|\boldsymbol{\Sigma}^{\top}\xi\right\|^2 (t-s)\right] \\
&= \exp\left[-\frac{1}{2}(t-s)\,\xi^{\top}\boldsymbol{\Sigma}\boldsymbol{\Sigma}^{\top}\xi\right]
\end{aligned}$$

which is the characteristic function of a $N\left(\mathbf{0}, (t-s)\,\boldsymbol{\Sigma}\boldsymbol{\Sigma}^{\top}\right)$ random vector, with $\boldsymbol{\Sigma}\boldsymbol{\Sigma}^{\top}$ a symmetric and positive semidefinite matrix.

It only remains to prove that $\boldsymbol{\Sigma}\boldsymbol{\Sigma}^{\top}$ is the correlation matrix of $\mathbf{X}$:

$$\begin{aligned}
\operatorname{Corr}\left(X_t^{(i)}, X_t^{(j)}\right) &= \frac{\operatorname{Cov}\left(X_t^{(i)}, X_t^{(j)}\right)}{\sqrt{\operatorname{Var}\left(X_t^{(i)}\right)}\sqrt{\operatorname{Var}\left(X_t^{(j)}\right)}} \\
&= \frac{\operatorname{Cov}\left(\sum_{k=1}^{d}\sigma_{ik}W_t^{(k)}, \sum_{l=1}^{d}\sigma_{jl}W_t^{(l)}\right)}{\sqrt{\operatorname{Var}\left(\sum_{k=1}^{d}\sigma_{ik}W_t^{(k)}\right)}\sqrt{\operatorname{Var}\left(\sum_{l=1}^{d}\sigma_{jl}W_t^{(l)}\right)}} \\
&= \frac{t\sum_{k=1}^{d}\sigma_{ik}\sigma_{jk}}{\sqrt{t\left(\sum_{k=1}^{d}\sigma_{ik}^2\right)}\sqrt{t\left(\sum_{l=1}^{d}\sigma_{jl}^2\right)}} \\
&= \sum_{k=1}^{d}\sigma_{ik}\sigma_{jk} = \left(\boldsymbol{\Sigma}\boldsymbol{\Sigma}^{\top}\right)_{ij}
\end{aligned}$$

□

The following result shows that every correlated Brownian motion can be represented in distribution as a linear transformation of a multidimensional Brownian motion.

Proposition 1.59. *If* ${}^{\mathbf{Q}}\mathbf{W} = \left({}^{\mathbf{Q}}W^{(1)}, \ldots, {}^{\mathbf{Q}}W^{(d)}\right)^{\top}$ *is a correlated Brownian motion with correlation matrix* $\mathbf{Q}$*, then the process* $\mathbf{X} := \mathbf{\Sigma W}$*, with* $\mathbf{\Sigma} = \mathbf{Q}^{1/2}$ *and* $\mathbf{W}$ *a* d*-dimensional Brownian motion, is a correlated Brownian motion with the same distribution that* ${}^{\mathbf{Q}}\mathbf{W}$.

Proof. As $\mathbf{Q}$ is symmetric and positive semidefinite, it admits a unique symmetric positive semidefinite square root, $\mathbf{\Sigma} = \mathbf{Q}^{1/2}$, and then $\mathbf{Q} = \mathbf{\Sigma}^2 = \mathbf{\Sigma\Sigma}^{\top}$. The characteristic function of ${}^{\mathbf{Q}}\mathbf{W}_t - {}^{\mathbf{Q}}\mathbf{W}_s$ can be written as

$$\begin{aligned}
\mathbb{E}\left[\exp\left(i\left\langle \xi, {}^{\mathbf{Q}}\mathbf{W}_t - {}^{\mathbf{Q}}\mathbf{W}_s\right\rangle\right)\right] &= \exp\left[-\frac{1}{2}(t-s)\,\xi^{\top}\mathbf{Q}\xi\right] \\
&= \exp\left[-\frac{1}{2}(t-s)\,\xi^{\top}\mathbf{\Sigma\Sigma}^{\top}\xi\right] \\
&= \exp\left[-\frac{1}{2}(t-s)\left(\xi^{\top}\mathbf{\Sigma}\right)\mathbf{I}\left(\mathbf{\Sigma}^{\top}\xi\right)\right] \\
&= \mathbb{E}\left[\exp\left(i\left\langle \mathbf{\Sigma}^{\top}\xi, \mathbf{W}_t - \mathbf{W}_s\right\rangle\right)\right] \\
&= \mathbb{E}\left[\exp\left(i\left\langle \xi, \mathbf{\Sigma}\left(\mathbf{W}_t - \mathbf{W}_s\right)\right\rangle\right)\right]
\end{aligned}$$

where we have applied that ${}^{\mathbf{Q}}\mathbf{W}_t - {}^{\mathbf{Q}}\mathbf{W}_s \sim N(\mathbf{0}, (t-s)\mathbf{Q})$ and that $\mathbf{W}_t - \mathbf{W}_s \sim N(\mathbf{0}, (t-s)\mathbf{I})$. Thus, we have that ${}^{\mathbf{Q}}\mathbf{W}_t - {}^{\mathbf{Q}}\mathbf{W}_s$ and $\mathbf{\Sigma}(\mathbf{W}_t - \mathbf{W}_s)$ are equally distributed. Taking $s = 0$ we conclude the proof. □

Proposition 1.60. *The components* ${}^{\mathbf{Q}}W^{(i)}$ *of a correlated Brownian motion are one-dimensional Brownian motions.*

Proof. Conditions i), ii) and iv) of Definition 1.5 follow directly from the corresponding conditions for correlated Brownian motion. Condition iii) is a immediate consequence of iii)' of Definition 1.57 and the fact that the diagonal entries of the correlation matrix $\mathbf{Q}$ are equal to 1. □

We know from Theorem 1.25 that the quadratic variation of one-dimensional Brownian motion is given by $[W, W](T) = T$. In order to extend this concept to multiple Brownian motions, we need the following definition.

Definition 1.61. (Quadratic covariation) The quadratic covariation of two functions $f, g : [0, T] \to \mathbb{R}$, $[f, g](T)$, is defined as

$$[f, g](T) = \lim_{\|\Pi_n\| \to 0} \sum_{j=0}^{n-1} \left[f(t_{j+1}) - f(t_j)\right]\left[g(t_{j+1}) - g(t_j)\right]$$

where $\Pi_n = \{t_0, t_1, \ldots, t_n\}$ is a partition of $[0, T]$ and $\|\Pi_n\| = \max_{j \in \{0, \ldots, n-1\}} (t_{j+1} - t_j)$.

Remark 1.62. There is no need to introduce a different notation for quadratic covariation, since if $f = g$, the definitions of quadratic covariation and quadratic variation coincide.

The following result states some properties of quadratic covariation.

Proposition 1.63. *Let $f, g : [0,T] \to \mathbb{R}$ be two functions such that their quadratic variation is well defined, and consider the mapping $[f,g] : [0,T] \to \mathbb{R}$, defined by $t \mapsto [f,g](t)$. Then, the following properties hold:*

i) (Polarization identity)

$$[f,g](t) = \frac{1}{2}\left([f+g, f+g](t) - [f,f](t) - [g,g](t)\right)$$

ii) (Symmetry) $[f,g](t) = [g,f](t)$

iii) (Bilinearity) For any $\alpha_1, \alpha_2, \beta_1, \beta_2 \in \mathbb{R}$ and any $f_1, f_2, g_1, g_2 : [0,T] \to \mathbb{R}$:

$$\begin{aligned}[\alpha_1 f_1 + \beta_1 g_1, \alpha_2 f_2 + \beta_2 g_2](t) &= \alpha_1\alpha_2 [f_1, f_2](t) + \alpha_1\beta_2 [f_1, g_2](t) \\ &\quad + \beta_1\alpha_2 [g_1, f_2](t) + \beta_1\beta_2 [g_1, g_2](t)\end{aligned}$$

iv) (Finite variation) $[f,g]$ is of finite variation.

Proof. i) It suffices to consider the equality

$$\begin{aligned}\left[(f+g)(t_{j+1}) - (f+g)(t_j)\right]^2 &= \left[f(t_{j+1}) - f(t_j)\right]^2 + \left[g(t_{j+1}) - g(t_j)\right]^2 \\ &\quad + 2\left[f(t_{j+1}) - f(t_j)\right]\left[g(t_{j+1}) - g(t_j)\right]\end{aligned}$$

and Definitions 1.23 and 1.61.

ii) Immediate from i).

iii) It suffices to apply Definition 1.61 and the properties of basic operations with functions.

iv) By Definition 1.23, quadratic variation is increasing. Then, by Proposition 1.22, it is of finite variation, and so is the quadratic covariation by the polarization identity. □

Before obtaining the quadratic covariation of correlated Brownian motion, we present a general result that will be useful later.

Theorem 1.64. *Let $f, g : [0,T] \to \mathbb{R}$ be two functions. If f is of finite variation and g is continuous, then $[f,g](T) = 0$.*

Proof. By Definition 1.61 we have

$$\begin{aligned}[f,g](T) &= \lim_{\|\Pi_n\| \to 0} \sum_{j=0}^{n-1} \left[f(t_{j+1}) - f(t_j)\right]\left[g(t_{j+1}) - g(t_j)\right] \\ &\leq \lim_{\|\Pi_n\| \to 0} \max_j \left|g(t_{j+1}) - g(t_j)\right| \sum_{j=0}^{n-1} \left[f(t_{j+1}) - f(t_j)\right] \\ &\leq \lim_{\|\Pi_n\| \to 0} \max_j \left|g(t_{j+1}) - g(t_j)\right| \sum_{j=0}^{n-1} \left|f(t_{j+1}) - f(t_j)\right| \\ &= TV(f) \lim_{\|\Pi_n\| \to 0} \max_j \left|g(t_{j+1}) - g(t_j)\right|\end{aligned}$$

The continuity of g implies its uniform continuity on $[0,T]$, therefore

$$\lim_{\|\Pi_n\|\to 0} \max_j \left|g\left(t_{j+1}\right) - g\left(t_j\right)\right| = 0$$

and then, $[f,g]\,(T) = 0$. □

Theorem 1.65. *Let* ${}^{\mathbf{Q}}\mathbf{W} = \left({}^{\mathbf{Q}}W^{(1)}, \ldots, {}^{\mathbf{Q}}W^{(d)}\right)^{\top}$ *be a correlated Brownian motion with correlation matrix* $\mathbf{Q}$. *Then*

$$\left[{}^{\mathbf{Q}}W^{(i)}, {}^{\mathbf{Q}}W^{(j)}\right](T) = q_{ij}T \tag{1.12}$$

in L^2 *sense, where* $q_{ij} = (\mathbf{Q})_{ij}$.[10]

Proof. Define $C^{ij}_{\Pi_n} := \sum_{k=0}^{n-1}\left[{}^{\mathbf{Q}}W^{(i)}_{t_{k+1}} - {}^{\mathbf{Q}}W^{(i)}_{t_k}\right]\left[{}^{\mathbf{Q}}W^{(j)}_{t_{k+1}} - {}^{\mathbf{Q}}W^{(j)}_{t_k}\right]$. We need to prove that $\lim_{\|\Pi_n\|\to 0}\mathbb{E}\left[\left(C^{ij}_{\Pi_n} - q_{ij}T\right)^2\right] = 0$. By the definition of $C^{ij}_{\Pi_n}$ we have

$$\begin{aligned} C^{ij}_{\Pi_n} - q_{ij}T &= \sum_{k=0}^{n-1}\left\{\left[{}^{\mathbf{Q}}W^{(i)}_{t_{k+1}} - {}^{\mathbf{Q}}W^{(i)}_{t_k}\right]\left[{}^{\mathbf{Q}}W^{(j)}_{t_{k+1}} - {}^{\mathbf{Q}}W^{(j)}_{t_k}\right] - q_{ij}\left(t_{k+1} - t_k\right)\right\} \\ &:= \sum_{k=0}^{n-1} Y^{ij}_k \end{aligned}$$

The variables Y^{ij}_k, $k = 0, \ldots, n-1$ are independent since the increments of correlated Brownian motion are independent by its definition. On the other hand

$$\begin{aligned} \mathbb{E}\left[Y^{ij}_k\right] &= \operatorname{Cov}\left\{\left[{}^{\mathbf{Q}}W^{(i)}_{t_{k+1}} - {}^{\mathbf{Q}}W^{(i)}_{t_k}\right], \left[{}^{\mathbf{Q}}W^{(j)}_{t_{k+1}} - {}^{\mathbf{Q}}W^{(j)}_{t_k}\right]\right\} \\ &\quad + \mathbb{E}\left[{}^{\mathbf{Q}}W^{(i)}_{t_{k+1}} - {}^{\mathbf{Q}}W^{(i)}_{t_k}\right]\mathbb{E}\left[{}^{\mathbf{Q}}W^{(j)}_{t_{k+1}} - {}^{\mathbf{Q}}W^{(j)}_{t_k}\right] - q_{ij}\delta_{ij}\left(t_{k+1} - t_k\right) \\ &= q_{ij}\left(t_{k+1} - t_k\right) - q_{ij}\left(t_{k+1} - t_k\right) = 0 \end{aligned}$$

where we have applied iii)' of Definition 1.57.

---

[10] The extension of equation (1.12) to a continuum of correlated Brownian motions give rise to the so-called *stochastic string models* (Santa-Clara and Sornette, 2001; Bueno-Guerrero et al., 2015).

Therefore

$$\mathbb{E}\left[\left(C^{ij}_{\Pi_n} - q_{ij}T\right)^2\right] = \mathbb{E}\left[\left(\sum_{k=0}^{n-1} Y_k^{ij}\right)^2\right] = \text{var}\left[\sum_{k=0}^{n-1} Y_k^{ij}\right] = \sum_{k=0}^{n-1} \text{var}\left[Y_k^{ij}\right]$$

$$= \sum_{k=0}^{n-1} \mathbb{E}\left[\left(Y_k^{ij}\right)^2\right]$$

$$= \sum_{k=0}^{n-1} \mathbb{E}\left\{\left[{}^{\mathbf{Q}}W^{(i)}_{t_{k+1}} - {}^{\mathbf{Q}}W^{(i)}_{t_k}\right]\left[{}^{\mathbf{Q}}W^{(j)}_{t_{k+1}} - {}^{\mathbf{Q}}W^{(j)}_{t_k}\right]\right.$$

$$\left. - q_{ij}\left(t_{k+1} - t_k\right)\right\}^2$$

$$= \sum_{k=0}^{n-1} \mathbb{E}\left\{\left[{}^{\mathbf{Q}}W^{(i)}_{t_{k+1}} - {}^{\mathbf{Q}}W^{(i)}_{t_k}\right]^2\left[{}^{\mathbf{Q}}W^{(j)}_{t_{k+1}} - {}^{\mathbf{Q}}W^{(j)}_{t_k}\right]^2\right.$$

$$- 2q_{ij}\left(t_{k+1} - t_k\right)\left[{}^{\mathbf{Q}}W^{(i)}_{t_{k+1}} - {}^{\mathbf{Q}}W^{(i)}_{t_k}\right]\left[{}^{\mathbf{Q}}W^{(j)}_{t_{k+1}} - {}^{\mathbf{Q}}W^{(j)}_{t_k}\right]$$

$$\left. + q_{ij}^2\left(t_{k+1} - t_k\right)^2\right\}$$

$$= \sum_{k=0}^{n-1} \mathbb{E}\left\{\left[{}^{\mathbf{Q}}W^{(i)}_{t_{k+1}} - {}^{\mathbf{Q}}W^{(i)}_{t_k}\right]^2\right\}\mathbb{E}\left\{\left[{}^{\mathbf{Q}}W^{(j)}_{t_{k+1}} - {}^{\mathbf{Q}}W^{(j)}_{t_k}\right]^2\right\}$$

$$+ 2\left(t_{k+1} - t_k\right)^2 q_{ij}^2 - 2\left(t_{k+1} - t_k\right)^2 q_{ij}^2 + q_{ij}^2\left(t_{k+1} - t_k\right)^2$$

$$= \sum_{k=0}^{n-1} \left(t_{k+1} - t_k\right)^2 q_{ii}q_{jj} + q_{ij}^2\left(t_{k+1} - t_k\right)^2$$

$$= \left(1 + q_{ij}^2\right)\sum_{k=0}^{n-1}\left(t_{k+1} - t_k\right)^2$$

$$\leq \left(1 + q_{ij}^2\right)\|\Pi_n\|\sum_{k=0}^{n-1}\left(t_{k+1} - t_k\right)$$

$$= \left(1 + q_{ij}^2\right)\|\Pi_n\| T$$

where we have used that for bidimensional Gaussian random variables (X, Y), with zero expectation, it is satisfied that $\mathbb{E}\left[X^2Y^2\right] = \mathbb{E}\left[X^2\right]\mathbb{E}\left[Y^2\right] + 2\text{Cov}^2(X, Y)$, and that $q_{ii} = q_{jj} = 1$. Taking limits we finally arrive at

$$\lim_{\|\Pi_n\|\to 0} \mathbb{E}\left[\left(C^{ij}_{\Pi_n} - q_{ij}T\right)^2\right] = 0$$

which was the desired result. □

Remark 1.66. As in the one-dimensional case (see Remark 1.26), L^2-convergence implies convergence in probability. Moreover, under additional assumptions on the partition, almost sure convergence also holds.

Theorem 1.65 allows us to obtain the quadratic covariation of multidimensional Brownian motion.

Corollary 1.67. *If* $\mathbf{W} = \left(W^{(1)}, \ldots, W^{(d)}\right)^{\top}$ *is a multidimensional Brownian motion, then*

$$\left[W^{(i)}, W^{(j)}\right](T) = \delta_{ij} T$$

Proof. It suffices to consider $\mathbf{Q} = \mathbf{I}$, with $\mathbf{I}$ the identity matrix of order d. □

It is worth noting that both Theorem 1.65 and Corollary 1.67 concerning quadratic covariation are consistent with the quadratic variation of one-dimensional Brownian motion (see Theorem 1.25).

1.12 Solutions to the Exercises

Exercise 1.13

From Proposition 1.12 we know that X_t is a martingale when $b = -\frac{a^2}{2}$. On the other hand

$$\begin{aligned}\mathbb{E}\left[X_t \,|\mathcal{F}_s\right] &= \mathbb{E}\left[e^{aW_t+bt} \,|\mathcal{F}_s\right] = \mathbb{E}\left[e^{a(W_t-W_s)}e^{aW_s+bt} \,|\mathcal{F}_s\right] \\ &= e^{aW_s+bt}\mathbb{E}\left[e^{a(W_t-W_s)}\right] = e^{aW_s+bt}e^{a^2(t-s)/2} \\ &= e^{aW_s+bt+a^2(t-s)/2}.\end{aligned}$$

where we have applied that Brownian motion increments are independent of the past, that W_t is an adapted process, and the moment generating function of $W_t - W_s$. Thus, $\mathbb{E}\left[X_t \,|\mathcal{F}_s\right]$ is equal to e^{aW_s+bs} for all $t > s$ if, and only if, $b = -\frac{a^2}{2}$.

Exercise 1.14

From Proposition 1.12 we know that $e^{uW_t - \frac{u^2}{2}t}$ is a martingale. Thus, for $0 \leq s \leq t$

$$e^{uW_s - \frac{u^2}{2}s} = \mathbb{E}\left[e^{uW_t - \frac{u^2}{2}t} \,|\mathcal{F}_s\right] = e^{-\frac{u^2}{2}t}\mathbb{E}\left[e^{uW_t} \,|\mathcal{F}_s\right]$$

Taking $s = 0$ we obtain

$$\mathbb{E}\left[e^{uW_t}\right] = e^{\frac{u^2}{2}t}$$

Exercise 1.16

Let $s, t \geq 0$ such that $s \leq t$. The covariance function of X_t can be written as

$$\begin{aligned}c(s,t) &= \operatorname{Cov}(X_s, X_t) \\ &= \operatorname{Cov}\left(W_s^2 - s, W_t^2 - t\right) = \operatorname{Cov}\left(W_s^2, W_t^2\right) \\ &= \mathbb{E}\left[W_s^2 W_t^2\right] - \mathbb{E}\left[W_s^2\right]\mathbb{E}\left[W_t^2\right]\end{aligned} \tag{1.13}$$

For the first term in the RHS we have

$$\begin{aligned} W_s^2 W_t^2 &= W_s^2 \left[(W_t - W_s) + W_s \right]^2 \\ &= W_s^2 \left[(W_t - W_s)^2 + W_s^2 + 2W_s (W_t - W_s) \right] \\ &= W_s^2 (W_t - W_s)^2 + W_s^4 + 2W_s^3 (W_t - W_s) \end{aligned}$$

and then

$$\begin{aligned} \mathbb{E}\left[W_s^2 W_t^2\right] &= \mathbb{E}\left[W_s^2\right] \mathbb{E}\left[(W_t - W_s)^2\right] + \mathbb{E}\left[W_s^4\right] + 2\mathbb{E}\left[W_s^3\right] \mathbb{E}\left[W_t - W_s\right] \\ &= s(t-s) + 3s^2 = 2s^2 + st \end{aligned}$$

where we have applied the independence of the increments of Brownian motion, and that $\mathbb{E}\left[W_s^4\right] = 3s^2$, since $W_s \sim N(0, s)$. On the other hand, the second term in the RHS is $\mathbb{E}\left[W_s^2\right]\mathbb{E}\left[W_t^2\right] = st$.

Replacing these results in (1.13), we obtain

$$c(s,t) = 2s^2 + st - st = 2s^2$$

Considering now $s > t$ we arrive at $c(s,t) = 2t^2$ and finally at

$$c(s,t) = 2(s \wedge t)^2$$

As we know from Proposition 1.12, the process X_t is a martingale. The previous result and Theorem 1.15 tell us that X_t is not a Brownian motion.

Exercise 1.17

$$\mu(t) = \mathbb{E}[X_t] = e^{-\alpha t/2} \mathbb{E}\left[W_{e^{\alpha t}}\right] = 0$$

$$\begin{aligned} c(s,t) = \mathrm{Cov}(X_s, X_t) = \mathbb{E}[X_s X_t] &= \mathbb{E}\left[e^{-\alpha s/2} W_{e^{\alpha s}} e^{-\alpha t/2} W_{e^{\alpha t}}\right] \\ &= e^{-\alpha(s+t)/2} \mathbb{E}\left[W_{e^{\alpha s}} W_{e^{\alpha t}}\right] \\ &= e^{-\alpha(s+t)/2} \mathrm{Cov}\left(W_{e^{\alpha s}}, W_{e^{\alpha t}}\right) \\ &= e^{-\alpha(s+t)/2} \left(e^{\alpha s} \wedge e^{\alpha t}\right) \\ &= e^{-\alpha(s+t)/2} e^{\alpha(s \wedge t)} = e^{-\alpha|t-s|/2} \end{aligned}$$

Exercise 1.19

a) X_t is a Brownian motion by the scaling property, i) of Proposition 1.18, with $c = \frac{1}{9}$.

b) Taking conditional expectations on the definition of Y_t and using the martingale property of Brownian motion, we have

$$\begin{aligned} \mathbb{E}[Y_t | \mathcal{F}_t] = \mathbb{E}[W_{3t} - W_t | \mathcal{F}_t] &= \mathbb{E}[W_{3t} | \mathcal{F}_t] - \mathbb{E}[W_t | \mathcal{F}_t] \\ &= W_t - W_t = 0 \end{aligned}$$

Thus, as $\mathbb{E}[Y_t | \mathcal{F}_t] \neq Y_t$, we have that Y_t is not adapted and therefore it is not a Brownian motion.

c) The covariance function of Z_t is given by

$$\operatorname{Cov}(Z_s, Z_t) = \operatorname{Cov}\left(\sqrt{s}W_1, \sqrt{t}W_1\right) = \sqrt{st}\operatorname{Cov}(W_1, W_1) = \sqrt{st}$$

Then, $\operatorname{Cov}(Z_s, Z_t) \neq s \wedge t$, and by Theorem 1.15 Z_t is not a Brownian motion.

Exercise 1.56

a) The process X_t inherits from W_t and $\widetilde{W}_t$ the properties of being adapted, continuous and having increments independent of the past. For the remaining property, iii) of Definition 1.5, applying the independence of W_t and $\widetilde{W}_t$ and the characteristic function of Gaussian random variables, we have

$$\begin{aligned}
&\mathbb{E}\{\exp[iu(X_t - X_s)]\} \\
&= \mathbb{E}\left\{\exp\left[iu\left(\rho(W_t - W_s) + \sqrt{1-\rho^2}\left(\widetilde{W}_t - \widetilde{W}_s\right)\right)\right]\right\} \\
&= \mathbb{E}\{\exp[iu\rho(W_t - W_s)]\}\,\mathbb{E}\left\{\exp\left[iu\sqrt{1-\rho^2}\left(\widetilde{W}_t - \widetilde{W}_s\right)\right]\right\} \\
&= e^{-u^2\rho^2(t-s)/2}e^{-u^2(1-\rho^2)(t-s)/2} = e^{-u^2(t-s)/2}
\end{aligned}$$

and thus, $X_t - X_s \sim N(0, (t-s))$.

b) The covariance of W_t and X_t is

$$\begin{aligned}
\operatorname{Cov}(W_t, X_t) &= \operatorname{Cov}\left(W_t, \rho W_t + \sqrt{1-\rho^2}\widetilde{W}_t\right) \\
&= \rho\operatorname{Cov}(W_t, W_t) + \sqrt{1-\rho^2}\operatorname{Cov}\left(W_t, \widetilde{W}_t\right) \\
&= \rho t
\end{aligned}$$

where we have used the independence of W_t and $\widetilde{W}_t$, and that $\operatorname{Cov}(W_s, W_t) = s \wedge t$. On the other hand

$$\operatorname{Var}(X_t) = \operatorname{Var}\left(\rho W_t + \sqrt{1-\rho^2}\widetilde{W}_t\right) = \rho^2\operatorname{Var}(W_t) + \left(1-\rho^2\right)\operatorname{Var}\left(\widetilde{W}_t\right) = t$$

Therefore

$$\operatorname{Corr}(W_t, X_t) = \frac{\operatorname{Cov}(W_t, X_t)}{\sqrt{\operatorname{Var}(W_t)}\sqrt{\operatorname{Var}(X_t)}} = \rho$$

1.13 Problems

1. Let W_t be a Brownian motion. Prove that

$$\mathbb{E}\left[W_t^{2k}\right] = (2k-1)!!t^k$$

for every positive integer k, where $(2k-1)!! = (2k-1)\cdot(2k-3)\cdots 3\cdot 1$.

2. Let W_t be a Brownian motion. Determine the values of $a \in \mathbb{R}$ for which the process $X_t = aW_t^3 - tW_t$ is a martingale.

3. Let W_t be a Brownian motion. Prove that the processes

$$V_t = tW_t - \int_0^t W_s ds$$

and

$$X_t = W_t^3 - 3\int_0^t W_s ds$$

are martingales.

4. Let $\mathbf{W}_t$ be a d-dimensional Brownian motion. Prove that $X_t = \frac{1}{d}\|\mathbf{W}_t\|^2 - t$ is a martingale.

5. Let $\mathbf{W}_t$ be a d-dimensional Brownian motion. Define the complex-valued process

$$M_t = \exp\left(i\langle \mathbf{u}, \mathbf{W}_t\rangle + \frac{1}{2}\|\mathbf{u}\|^2 t\right), \quad \mathbf{u} \in \mathbb{R}^d$$

Prove that the processes $R_t = \mathrm{Re}(M_t)$ and $I_t = \mathrm{Im}(M_t)$ are martingales.

1.14 Notes

A brief historical introduction to the concept of Brownian motion, with greater detail about the contributions of Brown, Einstein and Wiener, is presented in Bueno-Guerrero (2020), where additional bibliographical references can be consulted. An English translation of Einstein's original paper on Brownian motion can be found in Einstein (1956). Wiener's article, in which the first rigorous mathematical approach to Brownian motion was presented, is Wiener (1923). The physical approach to Brownian motion presented here is mainly based on Toral and Colet (2014). A similar description of Einstein's study of Brownian motion can be found in Evans (2013).

The terminology and definitions related to stochastic processes and Brownian motion differ considerably in the literature. Therefore, in this book, we have used various textbooks in an attempt to present the content in a consistent way. The main references used include Karatzas and Shreve (1991), Shreve (2004) and Klebaner (2005). Other useful references have been Rogers and Williams (1994), Protter (2005), Mörters and Peres (2010), Schilling and Partzsch (2012), Le Gall (2016) and Björk (2020).

The content of the section on total and quadratic variation is mainly based on Shreve (2004), whose presentation emphasizes intuition and pedagogical clarity. Readers interested in a more detailed and rigorous treatment are referred to Protter (2005). References on the fractal nature of Brownian motion paths are comparatively scarce. Most of what is presented here is based on Haussermann (1989) and Falconer (2003). The same applies to the topic of Brownian motion in higher dimensions. Our approach is based on Björk (2020).

Chapter 2

Stochastic Integration

Stochastic calculus can be viewed as an extension of ordinary calculus to the study of stochastic processes. Although more general frameworks exist, we will focus on a class of stochastic processes known as Itô processes, which use Brownian motion as a source of randomness.

To work in stochastic calculus, we will need to develop concepts analogous to those familiar from ordinary calculus: integrals, the chain rule, and differential equations. The corresponding concepts of stochastic calculus will be stochastic integrals, Itô's formula, and stochastic differential equations, respectively.

In this chapter, we will focus on the study of stochastic integrals and Itô's formula, leaving stochastic differential equations for the next chapter. We will also address two other topics of great theoretical and practical importance: Girsanov's theorem and the martingale representation theorem.

The chapter begins with Section 2.1, which addresses the need for a calculus for stochastic processes and the ineffectiveness of ordinary calculus techniques for this purpose. Section 2.2 is dedicated to the construction of the stochastic integral of progressively measurable processes, known as the Itô integral, as well as to establishing its properties. In addition, the concept of local martingale is introduced. Section 2.3 introduces an alternative definition of stochastic integral, the Stratonovich integral, and establishes its relationship with the Itô integral. Next, in Section 2.4, the main result of stochastic calculus, known as Itô's formula, is stated, and different forms in which it appears are presented. In the same section, the conditions under which a function of a Brownian motion is a martingale are studied, and the so-called Tanaka formulas are obtained. Section 2.5 studies the relationship between different absolutely continuous probability measures defined in the same space. Several results are obtained, which include Bayes' formula, Lévy's characterization of Brownian motion, and Girsanov's theorem. Another result of great practical importance is the martingale representation theorem, to which Section 2.6 is devoted. The solutions to the proposed exercises, a collection of additional problems, and the chapter notes are presented in Sections 2.7, 2.8, and 2.9, respectively.

2.1 Motivation

Consider a particle moving in a straight line through a fluid. This particle will be subjected to a friction force proportional to its velocity. The equation describing its motion is

$$\frac{dX_t}{dt} = -\alpha X_t$$

where X_t is the velocity of the particle at time t. If the particle is small enough, it will be affected by random microscopic forces, the same ones that give rise to Brownian motion. The effect of these forces can be introduced into the equation of motion as follows

$$\frac{dX_t}{dt} = -\alpha X_t + \sigma \xi_t \tag{2.1}$$

where σ is a diffusion coefficient and ξ_t is *white noise*, which represents the random forces acting on the particle. White noise can be interpreted as the derivative of Brownian motion in the sense of distributions.[1] Then, we can rewrite equation (2.1) as

$$dX_t = -\alpha X_t dt + \sigma dW_t \tag{2.2}$$

Equation (2.2) is known as the *Langevin equation*, since it was obtained by Paul Langevin in his study of Brownian motion. We will return to this equation in the next chapter.

Using the formal identity $\xi_t = \frac{dW_t}{dt}$ and using the Leibniz integral rule, it is not difficult to show that the solution of equation (2.2) is given by

$$X_t = e^{-\alpha t} X_0 + \sigma \int_0^t e^{-\alpha(t-s)} dW_s \tag{2.3}$$

In order to give a rigorous meaning to this solution we need to define properly the integral of the right hand side.

As a first attempt we could consider integrals of the type

$$\int_0^t f(s)\, dg(s)$$

where f and g are functions with the appropriate properties. In ordinary calculus, these integrals are known as *Riemann-Stieljes integrals*. In order to define them, consider the functions $f, g : [a, b] \to \mathbb{R}$ and let $\Pi = \{t_0 = a < t_1 < \cdots < t_n = b\}$ be a partition of $[a, b]$ and $\xi_j \in [t_j, t_{j+1}]$, $j = 0, \ldots, n-1$.

Definition 2.1. (Riemann-Stieljes integral) The function f is Riemann-Stieljes integrable with respect to g if the following limit exists

$$\lim_{\|\Pi\| \to 0} \sum_{j=0}^{n-1} f(\xi_j) \left[g(t_{j+1}) - g(t_j) \right]$$

---

[1] As we know from Chapter 1, Brownian motion is not differentiable at any point. Thus, the derivative $\frac{dW_t}{dt} = \xi_t$ should be understood in the sense of Schwartz distributions, i.e.

$$\int_{-\infty}^{+\infty} \xi_t f(t)\, dt = -\int_{-\infty}^{+\infty} W_t f'(t)\, dt$$

for any compact supported, smooth function f, and defining $W_t = W_{-t}$ for $t < 0$. See Kuo (2006) for a rigorous definition of white noise and its relation with stochastic integration.

independently of the choice of the points $\xi_j \in [t_j, t_{j+1}]$. If f is Riemann-Stieljes integrable, we define its Riemann-Stieljes integral as

$$\int_a^b f(s)\, dg(s) := \lim_{\|\Pi\|\to 0} \sum_{j=0}^{n-1} f(\xi_j)\left[g(t_{j+1}) - g(t_j)\right]$$

Note that if we take $g(s) = s$, we recover the Riemann integral. The property of Riemann-Stieljes integrals that interests us the most appears in the following result, which can be obtained from the Banach-Steinhaus theorem.

Proposition 2.2. *If* $\int_a^b f(s)\, dg(s)$ *exists for any* $f \in C[a,b]$, *then* g *is of finite variation in* $[a,b]$.

Proof. See Schilling and Partzsch (2012), Corollary A.41. □

According to Proposition 2.2, if we want our integral to be defined for continuous integrands (as is reasonable), the integrating function g must be of finite variation. Thus, since Brownian motion paths have unbounded variation (see Theorem 1.29), it is not possible to define a reasonable concept of stochastic integral by Riemann-Stieljes integration.

One possible way to circumvent the problem with the Riemann-Stieljes integral is to formally apply integration by parts as follows

$$\int_0^t f(s)\, dW_s = [f(s)\, W_s]_0^t - \int_0^t W_s df(s) = f(t)\, W_t - \int_0^t W_s df(s) \qquad (2.4)$$

where the integral in the right-hand side makes sense as a Riemann-Stieljes integral as long as f is of finite variation.

The type of integration of expression (2.4) can be made rigorous, and is known as the *Paley-Wiener-Zygmund integral* (Paley et al., 1933) (see Exercise 2.56 below). However, it cannot be used for integrands that depend on the Brownian motion W_s, which excludes this integral as a candidate for the stochastic integral we are seeking.

Having explored some unsuccessful approaches, in the next section we will address the construction of the stochastic integral. Our approach will not be the most general possible, but rather the most basic one that allows us to cover all the objectives of this book. In Section 2.9, we will provide details and references about more general definitions of the stochastic integral.

2.2 The Stochastic Integral

The first thing we need to know when constructing a stochastic integral with respect to Brownian motion is what type of processes we will consider as integrands. In some applications to finance, the integral will correspond to the position held in an asset at a given point in time. Since it is reasonable to obtain the value of this position based on the information available up to that point in time, we will consider adapted integrands. Specifically, if g_t is the integrand process, we will assume that g_t is $\mathcal{F}_t$-measurable for each $t \geq 0$.

In order to make it possible to interchange the expectation and the integral (a property that will be very useful in what follows), it is necessary to consider a more restricted class of processes known as *progressively measurable processes*.

Definition 2.3. (Progressively measurable process) A process X is progressively measurable if for each $t \geq 0$ the mapping $X : [0,t] \times \Omega \to \mathbb{R}$ is measurable with respect to the σ-algebra $\mathcal{B}([0,t]) \otimes \mathcal{F}_t$.

The following result allows us to gain more intuition about progressively measurable processes.

Lemma 2.4. *Any process adapted and right or left-continuous is also progressively measurable.*

Proof. See Mörters and Peres (2010), Lemma 7.2. □

Now, we will proceed to construct the stochastic integral, starting by defining the most basic integrands for which it can be defined.

Definition 2.5. (Progressively measurable step process) A progressively measurable step process is a process of the form

$$g_t(\omega) = \sum_{i=1}^{k} \xi_i(\omega)\, \mathbf{1}_{(t_i, t_{i+1}]}(t) \tag{2.5}$$

with $0 \leq t_1 \leq \cdots \leq t_{k+1}$, and ξ_i, $i = 1, \ldots, k$ are $\mathcal{F}_{t_i}$-measurable random variables.

Note that the name of the processes just defined is consistent since, by definition, they are adapted and left-continuous, and then, by Lemma 2.4, they are progressively measurable.

Definition 2.6. (Stochastic integral of a progressively measurable step process) The stochastic integral of a progressively measurable step process of the form (2.5) is defined as

$$\int_0^\infty g_t dW_t := \sum_{i=1}^{k} \xi_i \left(W_{t_{i+1}} - W_{t_i}\right) \tag{2.6}$$

Exercise 2.7. Obtain the value of the stochastic integral

$$\int_0^\infty \mathbf{1}_{(a,b]}(t) dW_t$$

for $0 \leq a < b$.

The next step in the construction of the stochastic integral is to determine the existence of a sequence of progressively measurable step processes, g_t^n, that converges in a certain sense to the progressively measurable process g_t, so that we can define

$$\int_0^\infty g_t dW_t := \lim_{n\to\infty} \int_0^\infty g_t^n dW_t \tag{2.7}$$

The following result, tells us that such a sequence exists under certain conditions, with convergence in the norm $\|\cdot\|_2$ determined by[2]

$$\|g\|_2^2 := \mathbb{E}\left[\int_0^\infty g_t^2 dt\right]$$

[2]The norm $\|\cdot\|_2$ is the 2-norm associated with the Hilbert space

$$L^2(\Omega \times [0,\infty), \mathcal{F} \otimes \mathcal{B}([0,\infty)), \mathbb{P} \times \lambda)$$

where λ is the Lebesgue measure (see AppendixA).

Lemma 2.8. *For every progressively measurable process g satisfying* $\mathbb{E}\left[\int_0^\infty g_t^2 dt\right] < \infty$, *we can find a sequence* g^n *of progressively measurable step processes such that* $\lim_{n\to\infty} \|g^n - g\|_2 = 0$.

Proof. See Mörters and Peres (2010), Lemma 7.3. □

Once we have stated the existence of the approximating sequence, for the stochastic integral to be well defined, we need to show that the limit on the right-hand side of expression (2.7) exists, and that it is independent of the choice of the sequence. To do this, we need some preliminary results. The first one is the *Itô isometry* for progressively measurable step processes.

Lemma 2.9. *Let g be a progressively measurable step process satisfying*

$$\mathbb{E}\left[\int_0^\infty g_t^2 dt\right] < \infty$$

Then

$$\mathbb{E}\left[\left(\int_0^\infty g_t dW_t\right)^2\right] = \mathbb{E}\left[\int_0^\infty g_t^2 dt\right]$$

Proof. Taking $g_t(\omega) = \sum_{i=1}^k \xi_i(\omega)\,\mathbf{1}_{(t_i,t_{i+1}]}(t)$, we can write

$$\begin{aligned}
\mathbb{E}\left[\left(\int_0^\infty g_t dW_t\right)^2\right] &= \mathbb{E}\left[\left(\sum_{i=1}^k \xi_i\left(W_{t_{i+1}} - W_{t_i}\right)\right)\left(\sum_{j=1}^k \xi_j\left(W_{t_{j+1}} - W_{t_j}\right)\right)\right] \\
&= \mathbb{E}\left[\sum_{i=1}^k \xi_i^2\left(W_{t_{i+1}} - W_{t_i}\right)^2\right] \\
&\quad + 2\mathbb{E}\left[\sum_{i<j} \xi_i\xi_j\left(W_{t_{i+1}} - W_{t_i}\right)\left(W_{t_{j+1}} - W_{t_j}\right)\right]
\end{aligned}$$

For the first expectation on the right-hand side we have

$$\begin{aligned}
\mathbb{E}\left[\sum_{i=1}^k \xi_i^2\left(W_{t_{i+1}} - W_{t_i}\right)^2\right] &= \sum_{i=1}^k \mathbb{E}\left[\mathbb{E}\left[\xi_i^2\left(\left(W_{t_{i+1}} - W_{t_i}\right)^2\right)|\mathcal{F}_{t_i}\right]\right] \\
&= \sum_{i=1}^k \mathbb{E}\left[\xi_i^2\mathbb{E}\left[\left(W_{t_{i+1}} - W_{t_i}\right)^2|\mathcal{F}_{t_i}\right]\right] \\
&= \sum_{i=1}^k \mathbb{E}\left[\xi_i^2\mathbb{E}\left[\left(W_{t_{i+1}} - W_{t_i}\right)^2\right]\right] \\
&= \sum_{i=1}^k \mathbb{E}\left[\xi_i^2\right]\left(t_{i+1} - t_i\right) \\
&= \mathbb{E}\left[\sum_{i=1}^k \xi_i^2\left(t_{i+1} - t_i\right)\right]
\end{aligned}$$

$$= \mathbb{E}\left[\int_0^\infty g_t^2 dt\right]$$

where we have applied the tower property of conditional expectations, that ξ_i is $\mathcal{F}_{t_i}$-measurable, that Brownian motion increments are independent of the past and, in the last step, that $g_t^2 = \sum_{i=1}^k \xi_i^2 \mathbf{1}_{(t_i,t_{i+1}]}(t)$.

Similarly, for the second expectation we have

$$\begin{aligned}
&\mathbb{E}\left[\sum_{i<j} \xi_i\xi_j \left(W_{t_{i+1}} - W_{t_i}\right)\left(W_{t_{j+1}} - W_{t_j}\right)\right] \\
&\quad = \sum_{i<j} \mathbb{E}\left[\mathbb{E}\left[\xi_i\xi_j \left(W_{t_{i+1}} - W_{t_i}\right)\left(W_{t_{j+1}} - W_{t_j}\right)|\mathcal{F}_{t_j}\right]\right] \\
&\quad = \sum_{i<j} \mathbb{E}\left[\xi_i\xi_j \left(W_{t_{i+1}} - W_{t_i}\right)\mathbb{E}\left[\left(W_{t_{j+1}} - W_{t_j}\right)|\mathcal{F}_{t_j}\right]\right] \\
&\quad = \sum_{i<j} \mathbb{E}\left[\xi_i\xi_j \left(W_{t_{i+1}} - W_{t_i}\right)\mathbb{E}\left[\left(W_{t_{j+1}} - W_{t_j}\right)\right]\right] = 0
\end{aligned}$$

where in the last step we have applied that Brownian motion increments have zero mean. □

Theorem 2.10. *Let g be a progressively measurable process satisfying*

$$\mathbb{E}\left[\int_0^\infty g_t^2 dt\right] < \infty$$

and g^n a sequence of progressively measurable step processes such that

$$\lim_{n\to\infty} \|g^n - g\|_2 = 0$$

Then, the limit

$$\lim_{n\to\infty} \int_0^\infty g_t^n dW_t \tag{2.8}$$

exists in L^2 sense and it is independent of the choice of g^n.

Proof. Using that $\lim_{n\to\infty} \|g^n - g\|_2 = 0$ and the triangle inequality, we can write

$$\begin{aligned}
0 = \lim_{n,m\to\infty} \|g^n - g^m\|_2^2 &= \lim_{n,m\to\infty} \mathbb{E}\left[\int_0^\infty \left(g_t^n - g_t^m\right)^2 dt\right] \\
&= \lim_{n,m\to\infty} \mathbb{E}\left[\left(\int_0^\infty \left(g_t^n - g_t^m\right) dW_t\right)^2\right]
\end{aligned}$$

where we have used that the difference of two progressively measurable step processes is again a progressively measurable step process, and Lemma 2.9. From the previous equality we conclude that $\int_0^\infty g_t^n dW_t$ is a Cauchy sequence in $L^2(\Omega, \mathcal{F}, \mathbb{P})$. As this space is complete, the limit (2.8) exists.

On the other hand, if g^n and h^n are two sequences of progressively measurable step processes satisfying $\lim_{n\to\infty} \|g^n - g\|_2 = \lim_{n\to\infty} \|h^n - g\|_2 = 0$, then by a similar reasoning, we obtain

$$0 = \lim_{n\to\infty} \|g^n - h^n\|_2^2 = \lim_{n\to\infty} \mathbb{E}\left[\left(\int_0^\infty (g_t^n - h_t^n)\, dW_t\right)^2\right]$$

from where we obtain that $\lim_{n\to\infty} \int_0^\infty g_t^n dW_t = \lim_{n\to\infty} \int_0^\infty h_t^n dW_t$ in L^2 sense. □

Theorem 2.10 guarantees that the following definition makes sense.

Definition 2.11. (Stochastic integral) Let g be a progressively measurable process satisfying $\mathbb{E}\left[\int_0^\infty g_t^2 dt\right] < \infty$ and let g^n be any sequence of progressively measurable step processes such that $\lim_{n\to\infty} \|g^n - g\|_2 = 0$. Then the stochastic integral of g is defined by the following limit in L^2 sense

$$\int_0^\infty g_t dW_t := \lim_{n\to\infty} \int_0^\infty g_t^n dW_t$$

So far, we have defined the stochastic integral as a random variable obtained from a progressively measurable process. For most applications, it will be important to transform the stochastic integral into an indefinite integral, thus converting it into a stochastic process.

Definition 2.12. (Stochastic integral as a stochastic process) Let g be a progressively measurable process satisfying $\mathbb{E}\left[\int_0^\infty g_t^2 dt\right] < \infty$. The stochastic integral of g as a process is defined as

$$\int_0^t g_s dW_s := \int_0^\infty \mathbf{1}_{s\le t} g_s dW_s$$

The stochastic integral as a process is well defined since $\mathbf{1}_{s\le t} g_s$ remains a progressively measurable process and satisfies

$$\mathbb{E}\left[\int_0^\infty (\mathbf{1}_{s\le t} g_s)^2 ds\right] = \mathbb{E}\left[\int_0^t g_s^2 ds\right] < \infty$$

From now on, we will refer to the stochastic integral as a process simply as stochastic integral.

The following result contains some properties of the stochastic integral.

Proposition 2.13. *Let g, h be two progressively measurable processes satisfying $\mathbb{E}\left[\int_0^\infty g_t^2 dt\right], \mathbb{E}\left[\int_0^\infty h_t^2 dt\right] < \infty$. Then, the following properties hold.*

i) (Linearity)

$$\int_0^t (a g_s + b h_s)\, dW_s = a \int_0^t g_s dW_s + b \int_0^t h_s dW_s \quad a, b \in \mathbb{R}$$

ii) (Itô isometry)

$$\mathbb{E}\left[\left(\int_0^t g_s dW_s\right)^2\right] = \mathbb{E}\left[\int_0^t g_s^2 ds\right]$$

iii) (Continuity) $\int_0^t g_s dW_s$ *has an almost surely continuous modification.*

iv) (Zero mean)

$$\mathbb{E}\left[\int_0^t g_s dW_s\right] = 0 \quad \forall t \geq 0$$

v) (Martingale) $\int_0^t g_s dW_s$ *is a martingale.*

vi) (Covariance)

$$\text{Cov}\left[\int_0^t g_s dW_s, \int_0^u h_s dW_s\right] = \mathbb{E}\left[\int_0^{t \wedge u} g_s h_s ds\right]$$

vii) (Quadratic covariation) If W_1 *and* W_2 *are two Brownian motions then*[3]

$$\left[\int_0^{\cdot} g_s dW_{1s}, \int_0^{\cdot} h_s dW_{2s}\right]_t = \int_0^t g_s h_s d\left[W_{1\cdot}, W_{2\cdot}\right]_s$$

viii) (Gaussian process) If g_s *is non-random, the stochastic integral* $\int_0^t g_s dW_s$ *is a Gaussian process with mean and covariance functions given, respectively, by* $\mu(t) = 0$ *and* $c(u, s) = \int_0^{u \wedge s} g_v^2 dv$.

Proof. i) Let g^n, h^n be sequences of progressively measurable processes satisfying $\lim_{n\to\infty} \|g^n - g\|_2 = \lim_{n\to\infty} \|h^n - h\|_2 = 0$. For the proof, it is enough to consider that $ag^n + bh^n$ is also a sequence of progressively measurable processes with $\lim_{n\to\infty} \|(ag^n + bh^n) - (ag + bh)\|_2 = 0$ and the linearity of the limit.

ii) Let $\overline{g}_s^n$ be a sequence of progressively measurable step processes converging to $\mathbf{1}_{s\leq t} g_s$. Then

$$\begin{aligned}
\mathbb{E}\left[\left(\int_0^t g_s dW_s\right)^2\right] &= \mathbb{E}\left[\left(\int_0^\infty \mathbf{1}_{s\leq t} g_s dW_s\right)^2\right] = \lim_{n\to\infty} \mathbb{E}\left[\left(\int_0^\infty \overline{g}_s^n dW_s\right)^2\right] \\
&= \lim_{n\to\infty} \mathbb{E}\left[\int_0^\infty \left(\overline{g}_s^n\right)^2 ds\right] = \mathbb{E}\left[\int_0^\infty \mathbf{1}_{s\leq t} g_s^2 ds\right] \\
&= \mathbb{E}\left[\int_0^t g_s^2 ds\right]
\end{aligned}$$

where we have applied Lemma 2.9 and Proposition B.50.

iii) For the proof of this result we refer the reader to Mörters and Peres (2010), Theorem 7.11 or Kallianpur and Sundar (2014), Section 5.2.

[3]From now on we adopt the standard notation $[X_\cdot, Y_\cdot]_t$ for the quadratic covariation of two stochastic processes. In Chapter 1, the equivalent notation $[X, Y](t)$ was used.

iv) Let $\overline{g}_s^n$ as in the proof of ii). By Theorem 2.10 and Definitions 2.11 and 2.12, we have

$$\lim_{n\to\infty} \mathbb{E}\left[\left(\int_0^\infty \overline{g}_s^n dW_s - \int_0^t g_s dW_s\right)^2\right] = 0$$

By Proposition B.50 we can interchange limit and expectation and write

$$\mathbb{E}\left[\int_0^t g_s dW_s\right] = \lim_{n\to\infty} \mathbb{E}\left[\int_0^\infty \overline{g}_s^n dW_s\right] = 0$$

where we have applied that $\int_0^\infty \overline{g}_s^n dW_s$ is of the form $\sum_{i=1}^{k_n} \xi_i \left(W_{t_{i+1}} - W_{t_i}\right)$, and then, $\mathbb{E}\left[\int_0^\infty \overline{g}_s^n dWs\right] = 0$.

v) By its definition, the stochastic integral is an adapted process. On the other hand, Itô isometry allows us to write

$$\begin{aligned}\mathbb{E}\left[\left(\int_0^t g_s dW_s\right)^2\right] &= \mathbb{E}\left[\left(\int_0^\infty \mathbf{1}_{s\le t} g_s dW_s\right)^2\right] = \mathbb{E}\left[\int_0^\infty \mathbf{1}_{s\le t} g_s^2 ds\right] \\ &\le \mathbb{E}\left[\int_0^\infty g_s^2 ds\right] < \infty\end{aligned}$$

and then, $\mathbb{E}\left[\left|\int_0^t g_s dW_s\right|\right] < \infty$. Finally, for every t, u with $0 \le u \le t$, by linearity, we have

$$\int_u^t g_s dW_s = \int_0^t [g_s - \mathbf{1}_{s\le u} g_s]\, dW_s = \int_0^t g_s dW_s - \int_0^u g_s dW_s$$

and thus

$$\begin{aligned}\mathbb{E}\left[\int_0^t g_s dW_s \,|\mathcal{F}_u\right] &= \mathbb{E}\left[\left(\int_0^u g_s dW_s + \int_u^t g_s dW_s\right) |\mathcal{F}_u\right] \\ &= \int_0^u g_s dW_s + \mathbb{E}\left[\int_u^t g_s dW_s \,|\mathcal{F}_u\right]\end{aligned}$$

By an argument analogous to that used in the proof of property iv), approximating g_s by progressively measurable step processes and passing to the limit, we obtain

$$\mathbb{E}\left[\int_u^t g_s dW_s \,|\mathcal{F}_u\right] = 0$$

This completes the proof of property v).

vi) Consider, without loss of generality, that $t \le u$. As the stochastic integral is a martingale with zero mean, we have

$$
\begin{aligned}
\mathrm{Cov}\left[\int_0^t g_s dW_s, \int_0^u h_s dW_s\right] &= \mathbb{E}\left[\left(\int_0^t g_s dW_s\right)\left(\int_0^u h_s dW_s\right)\right] \\
&= \mathbb{E}\left\{\mathbb{E}\left[\left(\int_0^t g_s dW_s\right)\left(\int_0^u h_s dW_s\right) | \mathcal{F}_t\right]\right\} \\
&= \mathbb{E}\left\{\left(\int_0^t g_s dW_s\right)\mathbb{E}\left[\left(\int_0^u h_s dW_s\right) | \mathcal{F}_t\right]\right\} \\
&= \mathbb{E}\left[\left(\int_0^t g_s dW_s\right)\left(\int_0^t h_s dW_s\right)\right] \\
&= \frac{1}{2}\mathbb{E}\left[\left(\int_0^t (g_s + h_s)\, dW_s\right)^2\right] - \frac{1}{2}\mathbb{E}\left[\left(\int_0^t g_s dW_s\right)^2\right] \\
&- \frac{1}{2}\mathbb{E}\left[\left(\int_0^t h_s dW_s\right)^2\right] \\
&= \frac{1}{2}\mathbb{E}\left[\int_0^t (g_s + h_s)^2\, ds\right] - \frac{1}{2}\mathbb{E}\left[\int_0^t g_s^2 ds\right] \\
&- \frac{1}{2}\mathbb{E}\left[\int_0^t h_s^2 ds\right] \\
&= \mathbb{E}\left[\int_0^t g_s h_s ds\right]
\end{aligned}
$$

vii) We refer the reader to Protter (2005), Theorem 29, page 74, where a general result for semimartingales is proven.

viii) Consider first the integral $\int_{t_j}^{t_{j+1}} g_s dW_s$. By construction, it is the L^2-limit of a sequence of linear combinations of Brownian motion increments with deterministic coefficients. Since each element of this sequence is a Gaussian random variable, and L^2-convergence implies convergence in distribution, the limit must also be Gaussian. Therefore, $\int_{t_j}^{t_{j+1}} g_s dW_s$ is a normal random variable. Moreover, by properties ii) and iv) we can state that

$$
\int_{t_j}^{t_{j+1}} g_s dW_s \sim N\left(0, \int_{t_j}^{t_{j+1}} g_s^2 ds\right) \tag{2.9}
$$

Now consider the processes $X_{t_j} = \int_0^{t_j} g_s dW_s$ for $j = 1, \ldots, k$ with $0 \le t_1 < \cdots < t_k$, and their linear combination

$$
Y = \sum_{j=1}^{k} a_j \int_0^{t_j} g_s dW_s = \sum_{j=0}^{k-1} c_j \int_{t_j}^{t_{j+1}} g_s dW_s
$$

for some $c_j \in \mathbb{R}$ and $t_0 = 0$.

The characteristic function of Y is given by

$$
\begin{aligned}
\mathbb{E}\left[e^{iuY}\right] &= \mathbb{E}\left\{\exp\left[iu\sum_{j=0}^{k-1} c_j \int_{t_j}^{t_{j+1}} g_s dW_s\right]\right\} \\
&= \mathbb{E}\left\{\prod_{j=0}^{k-1} \exp\left[iuc_j \int_{t_j}^{t_{j+1}} g_s dW_s\right]\right\} \\
&= \prod_{j=0}^{k-1} \mathbb{E}\left\{\exp\left[iuc_j \int_{t_j}^{t_{j+1}} g_s dW_s\right]\right\} \\
&= \prod_{j=0}^{k-1} \exp\left[-\frac{1}{2}u^2 c_j^2 \int_{t_j}^{t_{j+1}} g_s^2 ds\right] \\
&= \exp\left[-\frac{1}{2}u^2 \sum_{j=0}^{k-1} c_j^2 \int_{t_j}^{t_{j+1}} g_s^2 ds\right]
\end{aligned}
$$

where we have used that stochastic integrals with deterministic integrands over non-overlapping intervals are independent (a consequence of the independence of Brownian motion increments), and (2.9). Thus, Y has the characteristic function of a Gaussian random variable with zero mean and variance $\sum_{j=0}^{k-1} c_j^2 \int_{t_j}^{t_{j+1}} g_s^2 ds$. Therefore, we have proved that any linear combination of $\int_0^{t_1} g_s dW_s, \ldots, \int_0^{t_k} g_s dW_s$ is Gaussian. Then, the vector

$$
\left(\int_0^{t_1} g_s dW_s, \ldots, \int_0^{t_k} g_s dW_s\right)^{\top}
$$

is jointly Gaussian and $\int_0^t g_s dW_s$ is a Gaussian process.

Finally, by properties iv) and vi), we have

$$
\mu = \mathbb{E}\left[\int_0^t g_s dW_s\right] = 0
$$

and

$$
c(u,s) = \mathrm{Cov}\left[\int_0^u g_v dW_v, \int_0^s g_v dW_v\right] = \int_0^{u\wedge s} g_v^2 dv
$$

□

Some of the properties of the stochastic integral presented in Proposition 2.13 deserve further comment. Regarding Itô isometry, the following result provides a useful generalization valid for conditional expectations.

Proposition 2.14. ***(Conditional Itô isometry)*** *Let g be a progressively measurable process satisfying $\mathbb{E}\left[\int_0^\infty g_t^2 dt\right] < \infty$, and $0 \le s \le t$. Then*

$$
\mathbb{E}\left[\left(\int_0^t g_u dW_u\right)^2 | \mathcal{F}_s\right] = \mathbb{E}\left[\int_0^t g_u^2 du \,|\mathcal{F}_s\right] \tag{2.10}
$$

Proof. Let $A \in \mathcal{F}_s$ and consider the integrand $\overline{g}_u = \mathbf{1}_A g_u$. Applying Itô isometry to $\int_0^t \overline{g}_u dW_u$ we obtain

$$\mathbb{E}\left[\mathbf{1}_A \left(\int_0^t g_u dW_u\right)^2\right] = \mathbb{E}\left[\mathbf{1}_A \int_0^t g_u^2 du\right]$$

which is equivalent to (2.10). □

We know from Proposition 1.12 that $W_t^2 - t$ is a martingale. An analogous property holds for stochastic integrals.

Corollary 2.15. *If g is a progressively measurable process satisfying $\mathbb{E}\left[\int_0^\infty g_t^2 dt\right] < \infty$, then the process M_t defined by*

$$M_t = \left[\int_0^t g_u dW_u\right]^2 - \int_0^t g_u^2 du$$

is a martingale.

Proof. It is an immediate consequence of Proposition 2.14. □

According to property iii) of continuity, it is always possible to find a continuous modification of a stochastic integral. This will allow us to work with continuous versions, so from now on, we will consider stochastic integrals to be continuous processes.

The martingale property of stochastic integrals depends crucially on the integrability condition

$$\mathbb{E}\left[\int_0^\infty g_t^2 dt\right] < \infty \tag{2.11}$$

It is possible to define the stochastic integral under the weaker condition

$$\mathbb{P}\left[\int_0^\infty g_t^2 dt < \infty\right] = 1 \tag{2.12}$$

which allows to consider more general integrands. The construction of the integral is analogous to the one we have presented here, but with convergence in probability instead of L^2 sense. There also exists a continuous modification of the integral. However, the martingale property of the integral is lost, and we can only obtain, in general, the *local martingale* property (see, for example, Steele (2001), Chapter 7, for a more in-depth analysis, or Oksendal (1998), Chapter 3, and references therein).

Definition 2.16. (Local martingale, localizing sequence) The process X is a local martingale if there exists an increasing sequence τ_n of stopping times, with $\lim_{n\to\infty} \tau_n = \infty$ a.s., such that the processes $X_{t\wedge\tau_n}$ are martingales for each $n \geq 1$. We say that the sequence τ_n is a localizing sequence for the process X.[4]

Proposition 2.17. *Any martingale is a local martingale.*

---

[4]For each $n \in \mathbb{N}$, the process $X_{t\wedge\tau_n}$ is equal to X_t for $t < \tau_n$ and to a constant value X_{τ_n} for $t \geq \tau_n$. The process $X_{t\wedge\tau_n}$ is said to be the *process X stopped at τ_n*.

Proof. Let X be a martingale and consider the sequence of stopping times $\tau_n = n$, which is clearly increasing and satisfy that $\tau_n \longrightarrow \infty$ for $n \longrightarrow \infty$. The result follows by taking into account that a martingale stopped at a stopping time is also a martingale (see, for example, Le Gall (2016), Corollary 3.24). □

The converse of the previous proposition is not true in general. Local martingales that are not martingales do exist, although their construction typically requires techniques beyond the scope of this introductory book. For examples, see Durrett (1996), Example 2.1 and Exercise 2.2.

The condition $\mathbb{E}[|M_t|] < \infty$ for every $t \geq 0$ does not guarantee that a local martingale is a martingale. The examples cited above from Durrett (1996) provide counterexamples. A sufficient condition for the local martingale M_t to be a martingale is that $\mathbb{E}\left[\sup_{s\leq t}|M_s|\right] < \infty$ for every $t \geq 0$ (see Protter (2005), Theorem 51, page 38). In particular, a bounded local martingale is a martingale.

Returning to the extensions of Proposition 2.13, from property vii) we obtain the following consequences.

Corollary 2.18. *Let g, h be two progressively measurable processes satisfying*

$$\mathbb{E}\left[\int_0^\infty g_t^2 dt\right] < \infty$$

and

$$\mathbb{E}\left[\int_0^\infty h_t^2 dt\right] < \infty$$

Then, the following properties hold.

i) $\left[\int_0^\cdot g_s dW_s, \int_0^\cdot h_s dW_s\right]_t = \int_0^t g_s h_s ds$

ii) If ${}^Q\mathbf{W} = \left({}^QW_1, {}^QW_2\right)$ *is a two-dimensional correlated Brownian motion with correlation matrix*

$$Q = \begin{pmatrix} 1 & \rho \\ \rho & 1 \end{pmatrix}$$

then

$$\left[\int_0^\cdot g_s dW_{1s}, \int_0^\cdot h_s dW_{2s}\right]_t = \rho \int_0^t g_s h_s ds$$

iii) If W_1 and W_2 are two independent Brownian motions, then

$$\left[\int_0^\cdot g_s dW_{1s}, \int_0^\cdot h_s dW_{2s}\right]_t = 0$$

Proof. i) It suffices to apply $[W., W.]_t = t$.

ii) Immediate from Theorem 1.65.

iii) It is a direct consequence of Theorem 1.55 and Corollary 1.67. □

The following result states the relationship between the variance and the quadratic variation of the stochastic integral.

Proposition 2.19. *Under the conditions of Proposition 2.13 it is satisfied that*

$$\operatorname{Var}\left[\int_0^t g_s dW_s\right] = \mathbb{E}\left[\int_0^t g_s^2 ds\right] = \mathbb{E}\left\{\left[\int_0^\cdot g_s dW_s, \int_0^\cdot g_s dW_s\right]_t\right\}$$

Proof. It is an immediate consequence of properties vi) of Proposition 2.13 and i) of Corollary 2.18. □

Remark 2.20. Although condition (2.11) must be satisfied for the previous proposition to make sense, the expression

$$\left[\int_0^\cdot g_s dW_s, \int_0^\cdot g_s dW_s\right]_t = \int_0^t g_s^2 ds \tag{2.13}$$

is valid even under the weaker condition (2.12) (see, for example, Steele (2001), Theorem 8.6).

We already knew from Chapter 1 that Brownian motion has infinite variation. This property is inherited by stochastic integrals.

Corollary 2.21. *If $g_s \neq 0$ then $\int_0^t g_s dW_s$ is a process of infinite variation.*

Proof. An immediate consequence of iii) of Proposition 2.13, expression (2.13) and Theorem 1.28. □

Remark 2.22. Throughout this section the stochastic integral has been constructed as an L^2 random variable on $[0, \infty)$, following the approach of Schilling and Partzsch (2012). Consequently, the integrability condition is stated as

$$\mathbb{E}\left[\int_0^\infty g_s^2\, ds\right] < \infty.$$

In most applications, however, one works with the stochastic integral process of Definition 2.12 on a finite time interval. Since

$$\int_0^t g_s\, dW_s = \int_0^\infty \mathbf{1}_{s \le t} g_s\, dW_s,$$

it is enough to verify that

$$\mathbb{E}\left[\int_0^T g_s^2\, ds\right] < \infty$$

for the time horizon under consideration. For this reason, in the remainder of the book we shall usually check this latter condition.

2.3 Itô and Stratonovich Integrals

The consideration of integrands satisfying condition (2.12), and the convergence in probability allows us to obtain, under some conditions, the stochastic integral as the limit of Riemann sums.

Theorem 2.23. ***(Riemann representation)*** *Let g be a left-continuous progressively measurable process, and consider a partition* $\Pi = \{t_0, \ldots, t_n\}$ *of* $[0, t]$ *such that* $0 = t_0 < \cdots < t_n = t$. *Then*

$$\int_0^t g_s dW_s = \lim_{\|\Pi\| \to 0} \sum_{i=1}^n g_{t_{i-1}} \left(W_{t_i} - W_{t_{i-1}}\right) \tag{2.14}$$

where the limit is understood in the sense of convergence in probability.

Proof. See Revuz and Yor (1999), Proposition 2.13, Chapter IV. □

The fact that the value of the process evaluated at the lower end of each interval of the partition, $g_{t_{i-1}}$, appears in the Riemann sums of expression (2.14), is a distinctive feature of the type of stochastic integral that we are presenting here, known as *Itô integral* (Itô, 1944). However, there are other possible definitions of stochastic integral. To study them, we will use the result stated in the following exercise, taken from Karatzas and Shreve (1991) (Problem 2.29, Chapter 3).

Exercise 2.24. Let Π be a partition of $[0, t]$ with $0 = t_0 < \cdots < t_n = t$, and $\lambda \in [0, 1]$. Prove that

$$\lim_{\|\Pi\| \to 0} \sum_{i=1}^n \left[(1-\lambda) W_{t_{i-1}} + \lambda W_{t_i}\right] \left(W_{t_i} - W_{t_{i-1}}\right) = \frac{1}{2} W_t^2 + \left(\lambda - \frac{1}{2}\right) t \tag{2.15}$$

in L^2-sense.

If we take $\lambda = 0$ in (2.15), we obtain, by Theorem 2.23, the Itô integral (L^2 convergence implies convergence in probability)

$$\int_0^t W_s dW_s = \frac{1}{2} W_t^2 - \frac{1}{2} t \tag{2.16}$$

If instead of taking $\lambda = 0$, we take $\lambda = \dfrac{1}{2}$, we obtain the *Stratonovich integral*, which in this case is given by

$$\int_0^t W_s \circ dW_s = \frac{1}{2} W_t^2 \tag{2.17}$$

which is in agreement with the result we would obtain using ordinary calculus.[5]

Definition 2.25. (Stratonovich integral) Let g be a continuous and adapted process and $\Pi = \{t_0, \ldots, t_n\}$ a partition of $[0, t]$ with $0 = t_0 < \cdots < t_n = t$. The Stratonovich integral of g is defined as

$$\int_0^t g_s \circ dW_s = \frac{1}{2} \lim_{\|\Pi\| \to 0} \sum_{i=1}^n \left(g_{t_{i-1}} + g_{t_i}\right) \left(W_{t_i} - W_{t_{i-1}}\right)$$

with convergence in probability.

The following result states the relationship between Itô and Stratonovich integrals.

[5] Taking $\lambda = 1$ leads to the *backward Itô integral* (McKean (1969), page 35).

Proposition 2.26. *If g is a continuous and adapted process, then*

$$\int_0^t g_s \circ dW_s = \int_0^t g_s dW_s + \frac{1}{2} [g_\cdot, W_\cdot]_t \tag{2.18}$$

Proof. By Definition 2.25 we can write

$$\begin{aligned}
\int_0^t g_s \circ dW_s &= \frac{1}{2} \lim_{\|\Pi\|\to 0} \sum_{i=1}^{n} \left(g_{t_{i-1}} + g_{t_i}\right)\left(W_{t_i} - W_{t_{i-1}}\right) \\
&= \lim_{\|\Pi\|\to 0} \sum_{i=1}^{n} g_{t_{i-1}} \left(W_{t_i} - W_{t_{i-1}}\right) \\
&+ \frac{1}{2} \lim_{\|\Pi\|\to 0} \sum_{i=1}^{n} \left(g_{t_i} - g_{t_{i-1}}\right)\left(W_{t_i} - W_{t_{i-1}}\right) \\
&= \int_0^t g_s dW_s + \frac{1}{2} [g_\cdot, W_\cdot]_t
\end{aligned}$$

where we have applied Theorem 2.23 and Definition 1.61. □

In the case in which the integrand is a finite variation process, the two types of integrals coincide.

Corollary 2.27. *If g is a continuous and adapted process with finite variation, then*

$$\int_0^t g_s \circ dW_s = \int_0^t g_s dW_s$$

Proof. Immediate from Proposition 2.26 and Theorem 1.64. □

2.4 Itô's Formula

Let f be a continuously differentiable function, and h a continuous and increasing function on $[0, t]$. By the chain rule for Riemann-Stieljes integrals (see, for example, Schilling and Partzsch (2012), Theorem A.38), we have that

$$f(h(t)) = f(h(0)) + \int_0^t f'(h(s))\, dh(s) \tag{2.19}$$

In many of the applications that we will see later, we will need a formula, like the one above, for functions of the type $f(W_t)$, where W_t is a Brownian motion. However, as we already know (see expression (2.16)), the rules of ordinary calculus do not hold with Brownian motion calculus. Specifically, expression (2.19) is not applicable since W_t has no points of increase (see Section 1.7). Moreover, as Brownian motion has unbounded variation, it is possible that the integral of the right hand side does not even exist (Proposition 2.2).

As we will see below, the stochastic calculus formula analogous to (2.19), known as *Itô's formula*, *Itô's lemma* or *Itô's rule*, requires an additional term, which does not appear in ordinary calculus and which is a result of the infinite variation of Brownian motion in any interval.

Definition 2.28. (Itô process) An Itô process is a stochastic process of the form

$$X_t = X_0 + \int_0^t \mu_s ds + \int_0^t \sigma_s dW_s \quad 0 \le t \le T \tag{2.20}$$

where X_0 is $\mathcal{F}_0$-measurable, μ and σ are adapted processes, with σ progressively measurable, that satisfy $\int_0^T |\mu_t| \, dt < \infty$ and $\int_0^T \sigma_t^2 dt < \infty$ almost surely.

Remark 2.29. Note that the conditions on the processes μ and σ guarantee that the integrals in (2.20) are well defined.

Remark 2.30. The processes μ and σ in Definition 2.28 may depend on W_t, X_t or even on their past values up to time t. Most of the processes used in applications are Itô processes.

Recalling the motivation for a stochastic calculus from Section 2.1, it is often useful to express Itô processes in differential form. Thus, for example, we will express the process of expression (2.20) in the form

$$dX_t = \mu_t dt + \sigma_t dW_t \tag{2.21}$$

which is called the *stochastic differential* of the Itô process X. It is important to remark that expression (2.21) is formal, and it only makes sense in its integral form (2.20).

The following result will be useful in applications.

Lemma 2.31. *The integral $\int_0^t \mu_s ds$ in the definition of an Itô process is a finite variation process with zero quadratic variation.*

Proof. By Definition 2.28, $\int_0^t \mu_s ds$ is absolutely continuous in $[0, T]$ a.s. and therefore, it is of finite variation (Royden and Fitzpatrick (2010), Theorem 8, Chapter 6). Since absolute continuity implies continuity, by Theorem 1.28 the quadratic variation is zero. □

One of the advantages of Itô processes is that we can obtain immediately the quadratic covariation of two of them.

Theorem 2.32. *Let X, Y be two Itô processes of the form*

$$X_t = X_0 + \int_0^t \mu_s ds + \int_0^t \sigma_s dW_{1s}$$

$$Y_t = Y_0 + \int_0^t \lambda_s ds + \int_0^t \tau_s dW_{2s}$$

where W_1 and W_2 are two Brownian motions. Then, their quadratic covariation is given by

$$[X_\cdot, Y_\cdot]_t = \int_0^t \sigma_s \tau_s d\, [W_{1\cdot}, W_{2\cdot}]_s$$

or, in differential form

$$d\, [X_\cdot, Y_\cdot]_t = \sigma_t \tau_t d\, [W_{1\cdot}, W_{2\cdot}]_t$$

Proof. Applying the bilinearity of quadratic covariation, Lemma 2.31, the continuity of stochastic integrals, and Theorem 1.64, we have

$$[X_\cdot, Y_\cdot]_t = \left[\int_0^\cdot \sigma_s dW_{1s}, \int_0^\cdot \tau_s dW_{2s}\right]_t = \int_0^t \sigma_s \tau_s d\,[W_{1\cdot}, W_{2\cdot}]_s$$

by vii) of Proposition 2.13. □

Corollary 2.33. *If X is an Itô process of the form (2.20), then*

$$[X_\cdot, X_\cdot]_t = \int_0^t \sigma_s^2 ds$$

or

$$d\,[X_\cdot, X_\cdot]_t = \sigma_t^2 dt$$

Proof. It is an immediate consequence of Theorem 2.32. □

For a proper understanding of Itô's formula, which we present below, we need the following definition.

Definition 2.34. (Stochastic integral with respect to an Itô process) Let X be an Itô process given by

$$dX_t = \mu_t dt + \sigma_t dW_t \quad 0 \le t \le T \tag{2.22}$$

and h an adapted process such that $\int_0^T |h_t \mu_t|\, dt < \infty$ a.s. and $h\sigma$ is progressively measurable with $\int_0^T h_t^2 \sigma_t^2 dt < \infty$ a.s. Then, the stochastic integral of h with respect to X is defined as

$$\int_0^t h_s dX_s := \int_0^t h_s \mu_s ds + \int_0^t h_s \sigma_s dW_s$$

Remark 2.35. Note that the definition of $\int_0^t h_s dX_s$ can be obtained formally just by replacing dX_s by expression (2.22). Moreover, from the assumption on the process h, the integral is well defined and determines an Itô process. Using Theorem 2.32, it is also immediate to check that, if h, g are two integrand processes and X, Y are two Itô processes satisfying the conditions in Definition 2.34, then

$$\left[\int_0^\cdot g_s dX_s, \int_0^\cdot h_s dY_s\right]_t = \int_0^t g_s h_s d\,[X_\cdot, Y_\cdot]_s$$

We are now ready to state the stochastic version of (2.19), known as Itô's formula.

Theorem 2.36. ***(Itô's formula)*** *Let X be an Itô process and let $f : [0, \infty) \times \mathbb{R} \to \mathbb{R}$ be a function of class $C^{1,2}$. Then, the process $Y_t = f(t, X_t)$ has the stochastic differential*

$$df(t, X_t) = \frac{\partial f}{\partial t}(t, X_t)\, dt + \frac{1}{2}\frac{\partial^2 f}{\partial x^2}(t, X_t)\, d\,[X_\cdot, X_\cdot]_t + \frac{\partial f}{\partial x}(t, X_t)\, dX_t \tag{2.23}$$

Proof. See Lipster and Shiryaev (2001), page 124. □

Remark 2.37. If the Itô process X has the stochastic differential

$$dX_t = \mu_t dt + \sigma_t dW_t$$

then, by replacing it in Itô's formula (2.23) and using Corollary 2.33, we can rewrite Itô's formula in the following expanded form

$$\begin{aligned} df(t, X_t) &= \left[\frac{\partial f}{\partial t}(t, X_t) + \mu_t \frac{\partial f}{\partial x}(t, X_t) + \frac{1}{2}\sigma_t^2 \frac{\partial^2 f}{\partial x^2}(t, X_t) \right] dt \\ &\quad + \sigma_t \frac{\partial f}{\partial x}(t, X_t)\, dW_t \end{aligned} \tag{2.24}$$

In the simplest case in which $X_t = W_t$, we obtain the so-called *Itô's formula for Brownian motion*

$$df(t, W_t) = \left[\frac{\partial f}{\partial t}(t, W_t) + \frac{1}{2}\frac{\partial^2 f}{\partial x^2}(t, W_t) \right] dt + \frac{\partial f}{\partial x}(t, W_t)\, dW_t \tag{2.25}$$

Exercise 2.38. Obtain the stochastic differential of the process $X_t = e^{at+bW_t}$, with $a, b \in \mathbb{R}$.

Remark 2.39. We have presented Itô's formulas (2.23), (2.24) and (2.25) in their differential forms because they are the most commonly used in applications. The corresponding integral forms are

$$\begin{aligned} f(t, X_t) &= f(0, X_0) + \int_0^t \frac{\partial f}{\partial s}(s, X_s)\, ds + \frac{1}{2}\int_0^t \frac{\partial^2 f}{\partial x^2}(s, X_s)\, d[X_\cdot, X_\cdot]_s \\ &\quad + \int_0^t \frac{\partial f}{\partial x}(s, X_s)\, dX_s \end{aligned} \tag{2.26}$$

$$\begin{aligned} f(t, X_t) &= f(0, X_0) + \int_0^t \left[\frac{\partial f}{\partial s}(s, X_s) + \mu_s \frac{\partial f}{\partial x}(s, X_s) + \frac{1}{2}\sigma_s^2 \frac{\partial^2 f}{\partial x^2}(s, X_s) \right] ds \\ &\quad + \int_0^t \sigma_s \frac{\partial f}{\partial x}(s, X_s)\, dW_s \end{aligned}$$

$$f(t, W_t) = f(0, 0) + \int_0^t \left[\frac{\partial f}{\partial s}(s, W_s) + \frac{1}{2}\frac{\partial^2 f}{\partial x^2}(s, W_s) \right] ds + \int_0^t \frac{\partial f}{\partial x}(s, W_s)\, dW_s \tag{2.27}$$

Remark 2.40. If the function f does not depend explicitly on time t, Itô's formulas (2.23) and (2.24) can be rewritten, for $f : \mathbb{R} \to \mathbb{R}$ of class C^2, as

$$\begin{aligned} df(X_t) &= \frac{1}{2} f''(X_t)\, d[X_\cdot, X_\cdot]_t + f'(X_t)\, dX_t \\ &= \left[\mu_t f'(X_t) + \frac{1}{2}\sigma_t^2 f''(X_t) \right] dt + \sigma_t f'(X_t)\, dW_t \end{aligned} \tag{2.28}$$

or, in integral form

$$\begin{aligned} f(X_t) &= f(X_0) + \frac{1}{2}\int_0^t f''(X_s)\, d[X_\cdot, X_\cdot]_s + \int_0^t f'(X_s)\, dX_s \\ &= f(X_0) + \int_0^t \left[\mu_s f'(X_s) + \frac{1}{2}\sigma_s^2 f''(X_s)\right] ds \\ &+ \int_0^t \sigma_s f'(X_s)\, dW_s \end{aligned} \tag{2.29}$$

Exercise 2.41. Let g be a C^2 function. Obtain the value of $[g(W_\cdot), W_\cdot]_t$.

Exercise 2.42. Consider the Itô process given by

$$dX_t = \mu dt + \sigma dW_t, \quad X_0 = 0$$

with $\mu, \sigma \in \mathbb{R}$. Obtain $d\left(e^{\alpha X_t}\right)$ and $\mathbb{E}\left[e^{\alpha X_t}\right]$ with $\alpha \in \mathbb{R}$.

Exercise 2.43. Consider the Itô process given by

$$dX_t = \alpha X_t dt + \sigma X_t dW_t, \quad X_0 = 1 \tag{2.30}$$

where $\alpha, \sigma \in \mathbb{R}$. Obtain $\mathbb{E}[X_t]$ and $d\left(\frac{1}{X_t}\right)$.

Remark 2.44. In the conditions of the previous remark, if we take $dX_t = dW_t$, we obtain the time-independent version of Itô's formula for Brownian motion

$$df(W_t) = \frac{1}{2} f''(W_t)\, dt + f'(W_t)\, dW_t \tag{2.31}$$

or, in integral form

$$f(W_t) = f(0) + \frac{1}{2}\int_0^t f''(W_s)\, ds + \int_0^t f'(W_s)\, dW_s \tag{2.32}$$

Comparing this last formula with expression (2.19), we see how the chain rule of ordinary calculus has to be modified in order to be applied to stochastic calculus. We must add and an additional term $\frac{1}{2}\int_0^t f''(W_s)\, ds$ coming from the infinite variation of W, which leads to $[W_\cdot, W_\cdot]_t = t$. This term does not appear in (2.19) since h is continuous and has finite variation (as an increasing function), and then $[h(\cdot), h(\cdot)]_t = 0$ (Theorem 1.28).

As a first application of Itô's formula, if we know an antiderivative of the function f with respect to its second variable, it is possible to reformulate equation (2.25) in a way that is useful when obtaining stochastic integrals.

Corollary 2.45. *Let $F(t,x)$ be such that $f(t,x) = \frac{\partial F}{\partial x}(t,x)$, with $\frac{\partial F}{\partial t}$ and $\frac{\partial^2 F}{\partial x^2}$ continuous. Then*

$$\int_0^t f(s, W_s)\, dW_s = F(s, W_s)]_0^t - \int_0^t \left[\frac{\partial F}{\partial s}(s, W_s) + \frac{1}{2}\frac{\partial^2 F}{\partial x^2}(s, W_s)\right] ds$$

Proof. It suffices to replace f by F in expression (2.25). □

Exercise 2.46. Evaluate the following stochastic integrals:

a) $\int_0^t e^{W_s} dW_s$

b) $\int_0^t W_s e^{W_s} dW_s$

c) $\int_0^t e^{s/2} \cos W_s dW_s$

Another interesting application of Itô's formula, taken from Steele (2001), states sufficient conditions for a function of time and Brownian motion to be a local martingale or a martingale.

Proposition 2.47. ***(Martingale PDE condition)*** *If $f : [0, \infty) \times \mathbb{R} \to \mathbb{R}$ is of class $C^{1,2}$ and satisfies*

$$\frac{\partial f}{\partial t}(t,x) + \frac{1}{2}\frac{\partial^2 f}{\partial x^2}(t,x) = 0 \tag{2.33}$$

then $X_t = f(t, W_t)$ is a local martingale. If, in addition

$$\mathbb{E}\left\{\int_0^T \left[\frac{\partial f}{\partial x}(t, W_t)\right]^2 dt\right\} < \infty \tag{2.34}$$

then X_t is a martingale on $0 \le t \le T$.

Proof. Replacing (2.33) in the expression (2.27) of Itô's formula, we have

$$X_t = f(t, W_t) = f(0,0) + \int_0^t \frac{\partial f}{\partial x}(s, W_s)\, dW_s$$

As $\frac{\partial f}{\partial x}$ is continuous, the integral is well defined and is a local martingale. For the martingale property, it suffices to realize that (2.34) is actually condition (2.11). □

Itô's formula allows us to obtain stochastic differentials for smooth functions of Itô processes, but in some financial applications, we need to work with non-differentiable functions. For example, the process

$$[X_t - K]_+ := \max(X_t - K, 0) \tag{2.35}$$

where X_t is an Itô process and K some positive constant, can be written as $[X_t - K]_+ = f(X_t)$, with $f(x) = [x - K]_+ := \max(x - K, 0)$, which is a convex non-differentiable function.

The appropriate result for this type of functions is the following.

Theorem 2.48. ***(Generalized Itô's formula)*** *Let X be an Itô process and $f : \mathbb{R} \to \mathbb{R}$ a convex function.*[6] *Then*

$$f(X_t) = f(X_0) + \frac{1}{2}\int_{-\infty}^{+\infty} L_t^X(a)\, \mu(da) + \int_0^t D^- f(X_s)\, dX_s$$

[6] Theorem 2.48 is actually valid for any continous semimartingale X.

where $L_t^X(a)$ *is the local time of* X *at* a*,*[7] *defined by*

$$L_t^X(a) := \lim_{\varepsilon\to 0} \frac{1}{2\varepsilon}\int_0^t \mathbf{1}_{\{|X_s-a|\le\varepsilon\}} d[X_\cdot, X_\cdot]_s$$

and μ *is the second derivative measure, defined by*

$$\mu([a,b)) := D^- f(b) - D^- f(a)$$

where D^- *is the left-hand derivative*

$$D^- f(x) := \lim_{h\to 0^-} \frac{f(x+h)-f(x)}{h}$$

Proof. See Karatzas and Shreve (1991), Theorem 7.1, page 218 or Protter (2005), Corollary 3, page 225. □

Remark 2.49. Every convex function f has left-hand derivative $D^- f(x)$ for every $x \in \mathbb{R}$. Moreover, $D^- f(\cdot)$ is left-continuous (see Royden and Fitzpatrick (2010), Lemma 16, page 131 and Rockafellar (1970), Theorem 24.1).

Remark 2.50. If the function f is twice continuously differentiable, we have $D^- f(X_s) = f'(X_s)$, $\mu(da) = f''(a)\, da$ and

$$\int_{-\infty}^{+\infty} L_t^X(a)\,\mu(da) = \int_{-\infty}^{+\infty} f''(a)\, L_t^X(a)\, da = \int_0^t f''(X_s)\, d[X_\cdot, X_\cdot]_s$$

where we have applied the *occupation times formula*

$$\int_0^t g(X_s)\, d[X_\cdot, X_\cdot]_s = \int_{-\infty}^{+\infty} g(a)\, L_t^X(a)\, da \tag{2.36}$$

which is valid for any bounded measurable function g (see Klebaner (2005), Theorem 8.10, page 222 or Protter (2005), Corollary 1, page 216). Therefore, when f is of class C^2, we recover Itô's formula (2.29).

Remark 2.51. If we consider $X_t = W_t$, by the occupation times formula, with $g(x) = \mathbf{1}_A(x)$ and A a Borel set, we have

$$\int_0^t \mathbf{1}_A(W_s)\, ds = \int_{-\infty}^{+\infty} \mathbf{1}_A(a)\, L_t^W(a)\, da = \int_A L_t^W(a)\, da$$

The integral on the left-hand side is known as the *occupation time of Brownian motion on the set* A, and represents the time Brownian motion spends in the set A up to time t. Therefore, the previous equality allows us to interpret $L_t^W(a)$ as the time spent by Brownian motion in the infinitesimal interval $[a, a+da]$, and hence the name of (Brownian) local time. In the general case, where X is an Itô process, this interpretation is not so clear, since we have

$$\int_0^t \mathbf{1}_A(X_s)\, d[X_\cdot, X_\cdot]_s = \int_A L_t^X(a)\, da$$

and $L_t^X(a)$ can be considered as an occupation density, but relative to a "random clock", $[X_\cdot, X_\cdot]_t$ (Protter (2005), page 216).

[7] There are, in the literature, different definitions of $L_t^X(a)$ that differs in a factor 2. We have taken here the definition of Protter (2005).

We are now in a position to obtain the result equivalent to Itô's formula for the non-differentiable (but convex) functions $f(x) = [x-K]_+$ and $g(x) = |x-K|$. These formulas are known as *Tanaka-Meyer formulas.*

Corollary 2.52. ***(Tanaka-Meyer formulas)*** *If X is an Itô process, the following identities hold*

$$[X_t - K]_+ = [X_0 - K]_+ + \int_0^t \mathbf{1}_{(K,\infty)}(X_s)\, dX_s + \frac{1}{2} L_t^X(K)$$

$$|X_t - K| = |X_0 - K| + \int_0^t \operatorname{sgn}(X_s - K)\, dX_s + L_t^X(K)$$

where sgn *is the sign function given by*

$$\operatorname{sgn}(x) = \begin{cases} 1 & \text{if } x \geq 0 \\ -1 & \text{if } x < 0 \end{cases} \tag{2.37}$$

Proof. It is an immediate consequence of Theorem 2.48, taking into account that $D^- f(x) = \mathbf{1}_{(K,\infty)}(x)$ which implies $\mu(da) = \delta(a-K)\, da$ and $D^- g(x) = \mathbf{1}_{(0,\infty)}(x) - \mathbf{1}_{(-\infty,0]}(x) = \operatorname{sgn}(x)$ and then $\mu(da) = 2\delta(a-K)\, da$. □

Remark 2.53. It is possible to show that the local times $L_t^X(a)$ are adapted, continous and nondecreasing in t (Karatzas and Shreve (1991), pages 218-220).

Another extension of Theorem 2.36 consists of considering multidimensional Itô processes. We present it in the following result.

Theorem 2.54. ***(Multidimensional Itô's formula)*** *Let $\mathbf{X}_t = (X_{1t}, \ldots, X_{nt})^\top$ be a n-vector of Itô processes and let $f : [0,\infty) \times \mathbb{R}^n \to \mathbb{R}$ a $C^{1,2}$ function. Then, the process $f(t, \mathbf{X}_t)$ has the stochastic differential*

$$df(t,\mathbf{X}_t) = \frac{\partial f}{\partial t}(t,\mathbf{X}_t)\, dt + \frac{1}{2}\sum_{i=1}^n \sum_{j=1}^n \frac{\partial^2 f}{\partial x_i \partial x_j}(t,\mathbf{X}_t)\, d\left[X_{i\cdot}, X_{j\cdot}\right]_t + \sum_{i=1}^n \frac{\partial f}{\partial x_i}(t,\mathbf{X}_t)\, dX_{it} \tag{2.38}$$

Proof. See Le Gall (2016). □

The next result, known as *integration by parts* or *product rule*, allows us to obtain the differential of the product of two Itô processes.

Corollary 2.55. ***(Integration by parts)*** *If X,Y are two Itô processes, then*

$$d(X_t Y_t) = X_t dY_t + Y_t dX_t + d[X_\cdot, Y_\cdot]_t \tag{2.39}$$

Proof. It is enough to apply Theorem 2.54 to the function $f(x,y) = xy$. □

The following exercise shows how the Paley-Wiener-Zygmund integral of expression (2.4), fits perfectly into our presentation of the stochastic integral.

Exercise 2.56. Let f be an absolutely continuous function on $[0,T]$. Prove the Paley-Wiener-Zygmund integral

$$\int_0^t f(s)\, dW_s = f(t)\, W_t - \int_0^t W_s df(s) \qquad 0 < t \leq T$$

Exercise 2.57. Let X_t, Y_t be two Itô processes with $Y_t \neq 0$ a.s. $\forall t \geq 0$. Obtain the expression of $d\left(\frac{X_t}{Y_t}\right)$.

In both Itô's formula (2.23) and integration by parts formula (2.39), an extra term appears compared to the analogous formulas in ordinary calculus. We are already familiar with this characteristic of stochastic calculus with Itô integrals. Furthermore, we also know that the Stratonovich formulation of the stochastic integral allows us to recover some results from ordinary integration theory (see expression 2.17). This property of the Stratonovich integral also applies to Itô's formula (and consequently, to the integration by parts formula), which reduces to the chain rule of ordinary calculus.

In order to show this, we first need to define Stratonovich integrals with respect to Itô processes.

Definition 2.58. (Stratonovich integral with respect to an Itô process) If g is a continuous and adapted process, and X is an Itô process, the stochastic integral of g with respect to X is defined as

$$\int_0^t g_s \circ dX_s := \int_0^t g_s dX_s + \frac{1}{2}\left[g_\cdot, X_\cdot\right]_t$$

It is worth noting that this definition is consistent with expression (2.18) when the Itô process is a Brownian motion.

Theorem 2.59. ***(Itô's formula for Stratonovich calculus)*** *Let X be an Itô process and let $f : [0,\infty) \times \mathbb{R} \to \mathbb{R}$ be a function of class $C^{1,3}$. Then, the process $Y_t = f(t, X_t)$ has the following Stratonovich differential*

$$df(t, X_t) = \frac{\partial f}{\partial t}(t, X_t)\, dt + \frac{\partial f}{\partial x}(t, X_t) \circ dX_t \tag{2.40}$$

Proof. Consider the integral form of Itô's formula (2.26)

$$f(t, X_t) = f(0, X_0) + \int_0^t \frac{\partial f}{\partial s}(s, X_s)\, ds + \frac{1}{2}\int_0^t \frac{\partial^2 f}{\partial x^2}(s, X_s)\, d\left[X_\cdot, X_\cdot\right]_s + \int_0^t \frac{\partial f}{\partial x}(s, X_s)\, dX_s$$

By Definition 2.58 we can rewrite the last integral as

$$\int_0^t \frac{\partial f}{\partial x}(s, X_s)\, dX_s = \int_0^t \frac{\partial f}{\partial x}(s, X_s) \circ dX_s - \frac{1}{2}\left[\frac{\partial f}{\partial x}(\cdot, X_\cdot), X_\cdot\right]_t \tag{2.41}$$

In order to obtain the quadratic covariation on the right hand side, applying Itô's formula to $\frac{\partial f}{\partial x}(t, X_t)$ we have

$$\frac{\partial f}{\partial x}(t, X_t) = \frac{\partial f}{\partial x}(0, X_0) + A_t + \int_0^t \frac{\partial^2 f}{\partial x^2}(s, X_s)\, dX_s$$

where A_t collects the remaining finite variation terms. Then

$$\left[\frac{\partial f}{\partial x}(\cdot, X.), X.\right]_t = \left[\int_0^{\cdot} \frac{\partial^2 f}{\partial x^2}(s, X_s)\, dX_s, \int_0^{\cdot} dX_s\right]_t = \int_0^t \frac{\partial^2 f}{\partial x^2}(s, X_s)\, d\,[X., X.]_s$$

Replacing this result in (2.41) and then in (2.26), we obtain

$$f(t, X_t) = f(0, X_0) + \int_0^t \frac{\partial f}{\partial s}(s, X_s)\, ds + \int_0^t \frac{\partial f}{\partial x}(s, X_s) \circ dX_s$$

which is the integral form of equation (2.40). □

Similarly, the multidimensional Itô's formula can be expressed in Stratonovich form as

$$df(t, \mathbf{X}_t) = \frac{\partial f}{\partial t}(t, \mathbf{X}_t)\, dt + \sum_{i=1}^{n} \frac{\partial f}{\partial x_i}(t, \mathbf{X}_t) \circ dX_{it}$$

and, as a consequence, the integration by parts formula becomes

$$d(X_t Y_t) = X_t \circ dY_t + Y_t \circ dX_t$$

which is formally equal to its ordinary calculus counterpart.

As we have just seen, the Stratonovich formulation has the advantage, over Itô's formulation, of providing the same formulas as ordinary calculus. This makes it the natural formulation for working in certain fields, such as stochastic differential equations on manifolds (see Oksendal (1998) and references therein). On the other hand, the Stratonovich integral also has shortcomings, such as being defined for a narrower class of integrands and requiring more smoothness for Itô's formula.

However, the most important problem is that, unlike Itô's integral, which as we know is a martingale when condition (2.11) is satisfied, the Stratonovich integral is not (see expression (2.18)). The martingale (or local martingale) property of stochastic integrals is key to the applications of stochastic calculus in quantitative finance, as we will see later. Hence, in most situations, in financial applications, Itô's integral is used.

2.5 Girsanov's Theorem

In the applications of stochastic calculus to quantitative finance, there are situations in which different probability measures are used. One of the drawbacks arising from this situation is that some properties of stochastic processes, such as the martingale property, are not preserved when changing from one measure to another. Girsanov's theorem solves this problem by providing a method for obtaining martingales under the new measure from martingales under the original measure. We begin by defining the relevant types of probability measures and studying their properties.

Let $\mathbb{Q}$ be another probability measure defined on our probability space.

Definition 2.60. (Absolutely continuous and equivalent probability measures) The measure $\mathbb{Q}$ is absolutely continuous with respect to $\mathbb{P}$, $\mathbb{Q} << \mathbb{P}$, if $\mathbb{P}(A) = 0 \Rightarrow \mathbb{Q}(A) = 0$. If $\mathbb{Q} << \mathbb{P}$ and $\mathbb{P} << \mathbb{Q}$, then $\mathbb{P}$ and $\mathbb{Q}$ are equivalent probability measures, $\mathbb{P} \sim \mathbb{Q}$.

An immediate consequence of the previous definition is that equivalent probability measures share the same null sets, i.e., if $\mathbb{P} \sim \mathbb{Q}$ then $\mathbb{P}(A) = 0 \Leftrightarrow \mathbb{Q}(A) = 0$.

The following result tells us how we can obtain probabilities under $\mathbb{Q}$ by calculating expectations under $\mathbb{P}$, when $\mathbb{Q}$ is absolutely continous with respect to $\mathbb{P}$.

Theorem 2.61. ***(Radon-Nikodym)*** *If $\mathbb{Q} << \mathbb{P}$ then there exists a non-negative random variable Λ, satisfying $\mathbb{E}^{\mathbb{P}}[\Lambda] = 1$ and*

$$\mathbb{Q}(A) = \mathbb{E}^{\mathbb{P}}[\mathbf{1}_A \Lambda] = \int_A \Lambda d\mathbb{P} \tag{2.42}$$

for any $A \in \mathcal{F}$. The random variable Λ is unique $\mathbb{P}$-almost surely. [8]

Proof. This theorem is just a version of Theorem A.90 for probability measures, which are σ-finite by definition. □

Remark 2.62. The converse of the Radon-Nikodym Theorem is also true: If there exists a non-negative random variable Λ with $\mathbb{E}^{\mathbb{P}}[\Lambda] = 1$, then the set function $\mathbb{Q}$ defined by (2.42) determines a probability measure satisfying $\mathbb{Q} << \mathbb{P}$. The proof is an easy exercise in probability theory.

Remark 2.63. As $\mathbb{Q}$ is another probability measure, we have that $\mathbb{Q}(A) = \int_A d\mathbb{Q}$ $\forall A \in \mathcal{F}$. Considering this expression jointly with (2.42), we have that $d\mathbb{Q} = \Lambda d\mathbb{P}$, or $\Lambda = \frac{d\mathbb{Q}}{d\mathbb{P}}$, which gives rise to the following definition.

Definition 2.64. (Radon-Nikodym derivative) If $\mathbb{Q} << \mathbb{P}$, the random variable Λ of Theorem 2.61 is called the Radon-Nikodym derivative of $\mathbb{Q}$ with respect to $\mathbb{P}$, and it is denoted by $\frac{d\mathbb{Q}}{d\mathbb{P}}$.

We know from Theorem 2.61 that if $\mathbb{Q} << \mathbb{P}$ then $\frac{d\mathbb{Q}}{d\mathbb{P}} \geq 0$ a.s. The following result tells us that the equivalence of the measures guarantees that the Radon-Nikodym derivative is strictly positive.

Lemma 2.65. *If $\mathbb{Q} \sim \mathbb{P}$ then $\frac{d\mathbb{Q}}{d\mathbb{P}} > 0$ a.s.*

Proof. Let us assume, by contradiction, that there exists $B \in \mathcal{F}$ such that $\mathbb{P}(B) > 0$ and $\frac{d\mathbb{Q}}{d\mathbb{P}} = 0$ on B. Then $\mathbb{Q}(B) = \int_B \frac{d\mathbb{Q}}{d\mathbb{P}} d\mathbb{P} = 0$. But $\mathbb{Q} \sim \mathbb{P}$ and they must have the same null sets, which is a contradiction. □

Proposition 2.66. *If $\mathbb{Q} << \mathbb{P}$, for any random variable X integrable with respect to $\mathbb{Q}$, we have*

$$\mathbb{E}^{\mathbb{Q}}[X] = \mathbb{E}^{\mathbb{P}}[\Lambda X] \tag{2.43}$$

Proof. It suffices to consider the following chain of equalities

$$\mathbb{E}^{\mathbb{Q}}[X] = \int_\Omega X d\mathbb{Q} = \int_\Omega X \Lambda d\mathbb{P} = \mathbb{E}^{\mathbb{P}}[\Lambda X]$$

□

[8] $\mathbb{E}^{\mathbb{P}}$ represents mathematical expectation under the probability measure $\mathbb{P}$.

The Radon-Nikodym derivative possesses certain important properties, which thanks to the differential notation, can be written in a way reminiscent of analogous properties in ordinary calculus. The following results establish the chain rule and the derivative of the inverse function for Radon-Nikodym derivatives.

Proposition 2.67. *Let $\mathbb{Q}, \mathbb{P}, \mathbb{R}$ be probability measures on $(\Omega, \mathcal{F})$, such that $\mathbb{Q} << \mathbb{P}$ and $\mathbb{P} << \mathbb{R}$. Then $\mathbb{Q} << \mathbb{R}$ and*

$$\frac{d\mathbb{Q}}{d\mathbb{R}} = \frac{d\mathbb{Q}}{d\mathbb{P}}\frac{d\mathbb{P}}{d\mathbb{R}} \quad \mathbb{R}\text{-a.s.}$$

Proof. See Folland (1999), Proposition 3.9. □

Corollary 2.68. *If $\mathbb{Q} \sim \mathbb{P}$, then $\frac{d\mathbb{Q}}{d\mathbb{P}} = \left(\frac{d\mathbb{P}}{d\mathbb{Q}}\right)^{-1}$ a.s. respect to either $\mathbb{Q}$ or $\mathbb{P}$.*

Proof. It suffices to take $\mathbb{R} = \mathbb{P}$ in the previous proposition. □

One of the most important applications of stochastic calculus to quantitative finance is determining the conditional expectation of a random variable under one probability measure, given the conditional expectation under another equivalent probability measure and the Radon-Nikodym derivative of one measure with respect to the other. This is the content of the following result.

Theorem 2.69. ***(Bayes formula)*** *Let Λ be a positive $\mathbb{P}$-martingale with $\mathbb{E}^{\mathbb{P}}[\Lambda_T] = 1$. Consider the measure $\mathbb{Q}$ defined by $\frac{d\mathbb{Q}}{d\mathbb{P}} = \Lambda_T$. Then, if X is a $\mathcal{F}_t$-measurable random variable with $s \leq t \leq T$, and such that $X \in L^1(\Omega, \mathcal{F}, \mathbb{Q})$, we have*

$$\mathbb{E}^{\mathbb{Q}}[X|\mathcal{F}_s] = \frac{1}{\Lambda_s}\mathbb{E}^{\mathbb{P}}[\Lambda_t X|\mathcal{F}_s] = \frac{\mathbb{E}^{\mathbb{P}}[\Lambda_t X|\mathcal{F}_s]}{\mathbb{E}^{\mathbb{P}}[\Lambda_t|\mathcal{F}_s]} \quad \mathbb{Q}\text{-a.s.} \tag{2.44}$$

Proof. For any $A \in \mathcal{F}_s$ we have

$$\begin{aligned}
\mathbb{E}^{\mathbb{Q}}\left\{\mathbf{1}_A\frac{1}{\Lambda_s}\mathbb{E}^{\mathbb{P}}[\Lambda_t X|\mathcal{F}_s]\right\} &= \mathbb{E}^{\mathbb{P}}\left\{\mathbf{1}_A\frac{\Lambda_T}{\Lambda_s}\mathbb{E}^{\mathbb{P}}[\Lambda_t X|\mathcal{F}_s]\right\} \\
&= \mathbb{E}^{\mathbb{P}}\left\{\mathbb{E}^{\mathbb{P}}\left\{\mathbf{1}_A\frac{\Lambda_T}{\Lambda_s}\mathbb{E}^{\mathbb{P}}[\Lambda_t X|\mathcal{F}_s]\right\}|\mathcal{F}_s\right\} \\
&= \mathbb{E}^{\mathbb{P}}\left\{\mathbb{E}^{\mathbb{P}}\left\{\mathbf{1}_A\frac{\Lambda_T}{\Lambda_s}\mathbb{E}^{\mathbb{P}}[\Lambda_t X|\mathcal{F}_s]|\mathcal{F}_s\right\}\right\} \\
&= \mathbb{E}^{\mathbb{P}}\left\{\mathbf{1}_A\frac{1}{\Lambda_s}\mathbb{E}^{\mathbb{P}}[\Lambda_t X|\mathcal{F}_s]\,\mathbb{E}^{\mathbb{P}}[\Lambda_T|\mathcal{F}_s]\right\} \\
&= \mathbb{E}^{\mathbb{P}}\left\{\mathbf{1}_A\mathbb{E}^{\mathbb{P}}[\Lambda_t X|\mathcal{F}_s]\right\} = \mathbb{E}^{\mathbb{P}}[\mathbf{1}_A\Lambda_t X] \\
&= \mathbb{E}^{\mathbb{P}}\left[\mathbf{1}_A\mathbb{E}^{\mathbb{P}}[\Lambda_T|\mathcal{F}_t]\,X\right] = \mathbb{E}^{\mathbb{P}}[\mathbf{1}_A\Lambda_T X] \\
&= \mathbb{E}^{\mathbb{Q}}[\mathbf{1}_A X]
\end{aligned}$$

where we have used (2.43) with $\Lambda = \Lambda_T$, the tower property of conditional expectations, that X is $\mathcal{F}_t$-measurable, the martingale property of Λ under $\mathbb{P}$, and that

$A \in \mathcal{F}_s \subseteq \mathcal{F}_t$. Therefore, by the definition of conditional expectation (see Definition B.57), we have

$$\mathbb{E}^{\mathbb{Q}}[X|\mathcal{F}_s] = \frac{1}{\Lambda_s}\mathbb{E}^{\mathbb{P}}[\Lambda_t X|\mathcal{F}_s] \quad \mathbb{Q}\text{-a.s.}$$

which is expression (2.44). □

The first consequence of Theorem 2.69 is that the martingale property is not preserved under the change of measure. The relevant result on this regard is the following.

Corollary 2.70. *Under the assumptions of Theorem 2.69, a process M is a $\mathbb{Q}$-local martingale if, and only if, ΛM is a $\mathbb{P}$-local martingale.*

Proof. We first prove the result for martingales. If M is a martingale, by Theorem 2.69 we have

$$M_s = \mathbb{E}^{\mathbb{Q}}[M_t|\mathcal{F}_s] = \frac{1}{\Lambda_s}\mathbb{E}^{\mathbb{P}}[\Lambda_t M_t|\mathcal{F}_s]$$

and then $\mathbb{E}^{\mathbb{P}}[\Lambda_t M_t|\mathcal{F}_s] = \Lambda_s M_s$. For the integrability condition

$$\begin{aligned}
\mathbb{E}^{\mathbb{P}}[|\Lambda_t M_t|] = \mathbb{E}^{\mathbb{P}}[\Lambda_t|M_t|] &= \mathbb{E}^{\mathbb{P}}\left[\mathbb{E}^{\mathbb{P}}[\Lambda_T|\mathcal{F}_t]|M_t|\right] \\
&= \mathbb{E}^{\mathbb{P}}\left[\mathbb{E}^{\mathbb{P}}[\Lambda_T|M_t||\mathcal{F}_t]\right] = \mathbb{E}^{\mathbb{P}}[\Lambda_T|M_t|] \\
&= \mathbb{E}^{\mathbb{Q}}[|M_t|] < \infty
\end{aligned}$$

On the other hand, by (2.44)

$$\mathbb{E}^{\mathbb{Q}}[M_t|\mathcal{F}_s] = \frac{1}{\Lambda_s}\mathbb{E}^{\mathbb{P}}[\Lambda_t M_t|\mathcal{F}_s] = \frac{1}{\Lambda_s}\Lambda_s M_s = M_s$$

where we have applied that ΛM is a $\mathbb{P}$-martingale. Finally

$$\mathbb{E}^{\mathbb{Q}}[|M_t|] = \frac{1}{\Lambda_0}\mathbb{E}^{\mathbb{P}}[\Lambda_t|M_t|] = \mathbb{E}^{\mathbb{P}}[|\Lambda_t M_t|] < \infty$$

For local martingales, let $\{\tau_n\}_{n=1}^{\infty}$ be a localizing sequence for M. Then, $M_{t\wedge\tau_n}$ is a $\mathbb{Q}$-martingale for every n. By the martingale case proved above, $(\Lambda M)_{t\wedge\tau_n}$ is a $\mathbb{P}$-martingale for every n, and then, ΛM is a $\mathbb{P}$-local martingale. The converse implication follows by a similar reasoning. □

Not everything changes with the change of measure. Convergence in probability is preserved, as the following result states.

Theorem 2.71. *If $X_n \xrightarrow{\mathbb{P}} X$, and $\mathbb{Q} << \mathbb{P}$, then $X_n \xrightarrow{\mathbb{Q}} X$.*

Proof. See Klebaner (2005), Theorem 10.13. □

As a consequence of the previous result, the quadratic covariation is also preserved.

Corollary 2.72. *The quadratic covariation of two processes, obtained with respect to a probability measure does not change when obtained under another probability measure absolutely continuous with respect to the first.*[9]

Corollary 2.73. *Let X and Y be two stochastic processes and let $\mathbb{P}$ and $\mathbb{Q}$ be two probability measures such that $\mathbb{Q} \ll \mathbb{P}$. Then*

$$[X., Y.]_t^{\mathbb{Q}} = [X., Y.]_t^{\mathbb{P}}$$

for every $t \geq 0$, where the quadratic covariation is defined through convergence in probability.

Proof. By Definition 1.61, for any $t \leq T$,

$$\sum_{j=0}^{n-1} \left[X_{t_{j+1}} - X_{t_j}\right] \left[Y_{t_{j+1}} - Y_{t_j}\right] \underset{\|\Pi\| \to 0}{\overset{\mathbb{P}}{\longrightarrow}} [X., Y.]_t$$

and by Theorem 2.71, the same limit is obtained under $\mathbb{Q} << \mathbb{P}$. □

Before stating Girsanov's theorem we need to introduce some new concepts. The first one is the *stochastic exponential* or *Doléans-Dade exponential.*

Definition 2.74. (Stochastic exponential) Let X be an Itô process satisfying $X_0 = 0$ a.s. The stochastic exponential of X, $\mathcal{E}(X)$, is the stochastic process defined by

$$\mathcal{E}(X)_t := \exp\left\{X_t - \frac{1}{2}[X., X.]_t\right\}$$

Proposition 2.75. *The stochastic exponential of X has the stochastic differential*

$$d\mathcal{E}(X)_t = \mathcal{E}(X)_t \, dX_t \tag{2.45}$$

Proof. Making $V_t := X_t - \frac{1}{2}[X., X.]_t$ and $f(x) = e^x$, we have, by Itô's formula (2.28)

$$\begin{aligned} d\mathcal{E}(X)_t &= df(V_t) = \frac{1}{2} e^{V_t} d[V., V.]_t + e^{V_t} dV_t \\ &= \frac{1}{2}\mathcal{E}(X)_t \, d[X., X.]_t + \mathcal{E}(X)_t \left(dX_t - \frac{1}{2} d[X., X.]_t\right) \\ &= \mathcal{E}(X)_t \, dX_t \end{aligned}$$

where we have used that $[V., V.]_t = [X., X.]_t$ since $[X., X.]_t$ is a finite variation process and X is continuous. □

In ordinary calculus, the differential of the function $h(t) = e^{g(t)}$ is given by $dh(t) = h(t)\, dg(t)$. Thus, we can consider the stochastic exponential as the analogue, in stochastic calculus, of the ordinary exponential function. The difference with the

[9]In this corollary, we consider that the limit of the definition of quadratic covariance is understood with convergence in probability.

ordinary exponential lies in the presence of the term with $[X., X.]_t$, which is, in general, non-zero in the stochastic case.

If M is a continuous local martingale, then, by expression (2.45), $\mathcal{E}(M)_t$ can be represented as a stochastic integral with respect to a continuous local martingale. Therefore, $\mathcal{E}(M)$ is itself a continuous local martingale (see, for example, Revuz and Yor (1999), Proposition 2.10, Chapter IV). In stochastic calculus applications, it is sometimes useful to have a criterion that allows one to determine when the stochastic exponential of a local martingale is a true martingale. One of the most useful criteria is the so-called *Novikov condition.*

Theorem 2.76. ***(Novikov condition)*** *Let M be a continuous local martingale. If*

$$\mathbb{E}\left[\exp\left(\frac{1}{2}[M., M.]_t\right)\right] < \infty \quad \forall t \geq 0$$

then $\mathcal{E}(M)_t$ is a martingale.

Proof. See Karatzas and Shreve (1991), Proposition 5.12, Chapter 3. □

In order to prove Girsanov's theorem, we need two additional results. The first one is a characterization of Brownian motion obtained by Paul Lévy.

Theorem 2.77. ***(Lévy's characterization of Brownian motion)*** *The process M is a Brownian motion if, and only if, it is a continuous local martingale with $M_0 = 0$, and $[M., M.]_t = t$.*

Proof. We already know that Brownian motion W is a continuous local martingale with $W_0 = 0$ and $[W., W.]_t = t$. Let M be a continuous local martingale with $[M., M.]_t = t$. The process uM, with $u \in \mathbb{R}$, is also a continuous local martingale, and therefore, so is $\mathcal{E}(uM)_t$. Moreover

$$\mathbb{E}\left[\exp\left(\frac{1}{2}[uM., uM.]_t\right)\right] = e^{\frac{1}{2}u^2 t} < \infty \quad \forall t \geq 0$$

Then, by Novikov's condition $\mathcal{E}(uM)_t$ is a martingale. Applying this property we can write

$$\begin{aligned}
\mathbb{E}\left[e^{u(M_t - M_s)} \mid \mathcal{F}_s\right] &= \mathbb{E}\left[\mathcal{E}(uM)_t\, e^{\frac{1}{2}u^2 t} \mathcal{E}(-uM)_s\, e^{\frac{1}{2}u^2 s} \mid \mathcal{F}_s\right] \\
&= e^{\frac{1}{2}u^2 t + \frac{1}{2}u^2 s} \mathcal{E}(-uM)_s\, \mathcal{E}(uM)_s \\
&= e^{\frac{1}{2}u^2 t + \frac{1}{2}u^2 s} e^{-uM_s - \frac{1}{2}u^2 s} e^{uM_s - \frac{1}{2}u^2 s} \\
&= e^{\frac{1}{2}u^2 (t-s)}
\end{aligned}$$

Since the right hand side is non-random, M has increments independent of the past. Furthermore, taking expectations, we obtain the moment generating function of $M_t - M_s$ as

$$\mathbb{E}\left[e^{u(M_t - M_s)}\right] = e^{\frac{1}{2}u^2 (t-s)}$$

and thus, $M_t - M_s$ has a Gaussian distribution with zero mean and variance $(t - s)$. As M is adapted and continuous, we conclude that M is a Brownian motion. □

Exercise 2.78. Prove that $X_t = \int_0^t \operatorname{sgn}(W_s)\, dW_s$ is a Brownian motion.

As we know, the stochastic exponential of a continuous local martingale is itself another continuous local martingale. It is worth asking whether the converse is true: Is it possible to express any continuous local martingale as the stochastic exponential of another continuous local martingale? As the following result states, the answer is affirmative as long as the first local martingale is positive.

Proposition 2.79. *Let M be a positive continuous local martingale with $M_0 = 1$ a.s. Then, there exists a unique continuous local martingale N, such that $N_0 = 0$ a.s and $M_t = \mathcal{E}(N)_t$.*

Proof. By Itô's formula (2.29) with $f(x) = \ln x$ we have

$$\ln M_t = -\frac{1}{2}\int_0^t \frac{d[M.,M.]_s}{M_s^2} + \int_0^t \frac{dM_s}{M_s} \tag{2.46}$$

Defining

$$N_t := \int_0^t \frac{dM_s}{M_s}$$

we can rewrite (2.46) as

$$\ln M_t = N_t - \frac{1}{2}[N.,N.]_t$$

and we have proven that $M_t = \mathcal{E}(N)_t$ with $N_0 = 0$.

For uniqueness, let N and $\widetilde{N}$ be two continuous local martingales with $N_0 = \widetilde{N}_0 = 0$, and such that $M_t = \mathcal{E}(N)_t = \mathcal{E}\left(\widetilde{N}\right)_t$. Then, by the definition of the stochastic exponential we obtain

$$N_t - \widetilde{N}_t = \frac{1}{2}\left([N.,N.]_t - \left[\widetilde{N}.,\widetilde{N}.\right]_t\right)$$

Therefore, the continuous local martingale $N_t - \widetilde{N}_t$ is of finite variation and thus is almost surely constant (see Protter (2005), Corollary 1, page 72 and subsequent comments). As $N_0 = \widetilde{N}_0 = 0$, then $N_t - \widetilde{N}_t = 0$, and N_t is almost surely unique. □

In order to use Proposition 2.79 in the statement of Girsanov's theorem, we need to determine which local martingale (actually a martingale) will play the role of M in that proposition. We define that martingale below.

Definition 2.80. (Likelihood process) If $\mathbb{Q} << \mathbb{P}$, the process

$$L_t = \mathbb{E}^{\mathbb{P}}\left[\frac{d\mathbb{Q}}{d\mathbb{P}} | \mathcal{F}_t\right]$$

is called the likelihood process.

Now we are ready to state Girsanov's theorem.

Theorem 2.81. ***(Girsanov's theorem)*** *Let $W^{\mathbb{P}}$ be a Brownian motion under the probability measure $\mathbb{P}$. Let $\mathbb{Q}$ be another probability measure equivalent to $\mathbb{P}$, $\mathbb{Q} \sim \mathbb{P}$. Assume that the likelihood process, $L_t = \mathbb{E}^{\mathbb{P}}\left[\frac{d\mathbb{Q}}{d\mathbb{P}} | \mathcal{F}_t\right]$, is continuous, and consider the process N determined by $L_t = \mathcal{E}(N)_t$. Then, the process $W^{\mathbb{Q}}$ defined by*

$$W_t^{\mathbb{Q}} := W_t^{\mathbb{P}} - \left[W_\cdot^{\mathbb{P}}, N_\cdot\right]_t \tag{2.47}$$

is a $\mathbb{Q}$-Brownian motion.

Proof. First we will prove that the process N is well defined. By Lemma 2.65, we know that $\frac{d\mathbb{Q}}{d\mathbb{P}} > 0$ a.s., and then $L_t = \mathbb{E}^{\mathbb{P}}\left[\frac{d\mathbb{Q}}{d\mathbb{P}} | \mathcal{F}_t\right] > 0$ a.s. L_t is continuous by assumption. Let us see that it is also a $\mathbb{P}$-martingale. For $s < t$ we have

$$\mathbb{E}^{\mathbb{P}}\left[L_t | \mathcal{F}_s\right] = \mathbb{E}^{\mathbb{P}}\left[\mathbb{E}^{\mathbb{P}}\left[\frac{d\mathbb{Q}}{d\mathbb{P}} | \mathcal{F}_t\right] | \mathcal{F}_s\right] = \mathbb{E}^{\mathbb{P}}\left[\frac{d\mathbb{Q}}{d\mathbb{P}} | \mathcal{F}_s\right] = L_s$$

and

$$\mathbb{E}^{\mathbb{P}}\left[|L_t|\right] = \mathbb{E}^{\mathbb{P}}\left[L_t\right] = \mathbb{E}^{\mathbb{P}}\left[\mathbb{E}^{\mathbb{P}}\left[\frac{d\mathbb{Q}}{d\mathbb{P}} | \mathcal{F}_t\right]\right] = \mathbb{E}^{\mathbb{P}}\left[\frac{d\mathbb{Q}}{d\mathbb{P}}\right] = 1 < \infty$$

where we have used the Radon-Nikodym theorem. Then, L_t is a $\mathbb{P}$-martingale and, therefore, a $\mathbb{P}$-local martingale. Moreover, $L_0 = \mathbb{E}^{\mathbb{P}}\left[\frac{d\mathbb{Q}}{d\mathbb{P}}\right] = 1$. Being L_t a positive continuous local martingale with $L_0 = 1$, Proposition 2.79 guarantees the existence (and uniqueness) of the process N as a continuous local martingale satisfying $N_0 = 0$.

On the other hand, the process $\left[W_\cdot^{\mathbb{P}}, N_\cdot\right]_t$ is continuous (under $\mathbb{P}$) as it is the quadratic covariation of two continuous local martingales (Karatzas and Shreve (1991), Problem 5.17, Chapter 1). Thus, $W_t^{\mathbb{Q}} = W_t^{\mathbb{P}} - \left[W_\cdot^{\mathbb{P}}, N_\cdot\right]_t$ is a continuous process under $\mathbb{P}$. Since $\mathbb{Q} \sim \mathbb{P}$, $W_t^{\mathbb{Q}}$ is also continuous under $\mathbb{Q}$. Moreover, as $\left[W_\cdot^{\mathbb{P}}, N_\cdot\right]_t$ is of finite variation and $W_t^{\mathbb{P}}$ is continuous, we have $\left[W_\cdot^{\mathbb{Q}}, W_\cdot^{\mathbb{Q}}\right]_t = \left[W_\cdot^{\mathbb{P}}, W_\cdot^{\mathbb{P}}\right]_t = t$.

By Lévy's characterization of Brownian motion, it remains to prove that $W_t^{\mathbb{Q}}$ is a $\mathbb{Q}$-local martingale. By Corollary 2.70, this is equivalent to show that $L_t W_t^{\mathbb{Q}}$ is a $\mathbb{P}$-local martingale. Using integration by parts we have

$$d\left(L_t W_t^{\mathbb{Q}}\right) = L_t dW_t^{\mathbb{Q}} + W_t^{\mathbb{Q}} dL_t + d\left[L_\cdot, W_\cdot^{\mathbb{Q}}\right]_t \tag{2.48}$$

and by expression (2.45)

$$dL_t = L_t dN_t \tag{2.49}$$

On the other hand

$$d\left[L_\cdot, W_\cdot^{\mathbb{Q}}\right]_t = d\left[L_\cdot, W_\cdot^{\mathbb{P}}\right]_t = L_t d\left[N_\cdot, W_\cdot^{\mathbb{P}}\right]_t \tag{2.50}$$

where in the first step we have used that $\left[W_\cdot^{\mathbb{P}}, N_\cdot\right]_t$ is of finite variation and L_t is continuous, and in the second step, we have applied (2.49). Replacing (2.49) and (2.50) in (2.48), and using the definition of $W_t^{\mathbb{Q}}$, we have

$$\begin{aligned} d\left(L_t W_t^{\mathbb{Q}}\right) &= L_t\left(dW_t^{\mathbb{P}} - d\left[W_\cdot^{\mathbb{P}}, N_\cdot\right]_t\right) + W_t^{\mathbb{Q}} L_t dN_t + L_t d\left[N_\cdot, W_\cdot^{\mathbb{P}}\right]_t \\ &= L_t dW_t^{\mathbb{P}} + W_t^{\mathbb{Q}} L_t dN_t \end{aligned}$$

Thus, $L_t W_t^{\mathbb{Q}}$ can be written as the sum of two stochastic integrals with respect to continuous $\mathbb{P}$-local martingales ($W_t^{\mathbb{P}}$ and N_t), and then, it is a continuous $\mathbb{P}$-local martingale. □

Remark 2.82. If we know the explicit expression of N_t as a stochastic integral, $N_t = \int_0^t \theta_s dW_s^{\mathbb{P}}$, expression (2.47) takes the form

$$W_t^{\mathbb{Q}} = W_t^{\mathbb{P}} - \int_0^t \theta_s ds \tag{2.51}$$

Remark 2.83. We have stated Theorem 2.81 assuming that $\mathbb{P}$ and $\mathbb{Q}$ are known. There is an alternative version in which the measure $\mathbb{Q}$ is defined by its Radon-Nikodym derivative

$$\frac{d\mathbb{Q}}{d\mathbb{P}} := \mathcal{E}(X)_T = \exp\left\{\int_0^T \theta_s dW_s^{\mathbb{P}} - \frac{1}{2}\int_0^T \theta_s^2 ds\right\}$$

where $X_t := \int_0^t \theta_s dW_s^{\mathbb{P}}$, θ is such the integral is well defined, and we recover expression (2.51). This version of the theorem requires an additional assumption that guarantees that $\mathbb{E}^{\mathbb{P}}\left[\frac{d\mathbb{Q}}{d\mathbb{P}}\right] = 1$, since it is a necessary condition for $\frac{d\mathbb{Q}}{d\mathbb{P}}$ to define the probability measure $\mathbb{Q}$ (see Remark 2.62). Usually, this additional assumption is that $\mathcal{E}(X)_t$ is a martingale, since in this way we have

$$\mathbb{E}^{\mathbb{P}}\left[\frac{d\mathbb{Q}}{d\mathbb{P}}\right] = \mathbb{E}^{\mathbb{P}}\left[\mathcal{E}(X)_T \,|\mathcal{F}_0\right] = \mathcal{E}(X)_0 = 1$$

The martingale property of $\mathcal{E}(X)_t$ can be checked, for example, with Novikov's condition (Theorem 2.76).[10]

2.6 Martingale Representation

We know from Proposition 2.13 that stochastic integrals of progressively measurable processes satisfying condition (2.11) are martingales. In this section we will see that, under certain conditions, the inverse result is also true: given a martingale it is possible to express it as a stochastic integral. This result is known as Martingale Representation Theorem, and it is not only an interesting result in pure mathematics. It is a key tool in the study of market completeness.

We start the study of martingale representation with a technical result. Consider the set $\mathcal{H}_T^2$ of random variables of the form

$$M_T = x + \int_0^T g_s dW_s$$

with $x \in \mathbb{R}$ and g a progressively measurable process satisfying (2.11).

[10]It can be proved that $\mathcal{E}(X)_t$ is a *supermartingale*, i.e., $\mathbb{E}^{\mathbb{P}}\left[\mathcal{E}(X)_t \,|\mathcal{F}_s\right] \le \mathcal{E}(X)_s$, $s \le t$, and then $\mathbb{E}^{\mathbb{P}}\left[\frac{d\mathbb{Q}}{d\mathbb{P}}\right] \le 1$. Moreover, it is a martingale if, and only if, $\mathbb{E}^{\mathbb{P}}\left[\mathcal{E}(X)_t\right] = 1 \; \forall t \ge 0$ (Kallianpur and Sundar (2014), Lemma 5.7.1).

Lemma 2.84. *The space $\mathcal{H}_T^2$ is a closed linear subspace of $L^2(\Omega, \mathcal{F}_T, \mathbb{P})$.*

Proof. By Itô isometry we have that $\mathcal{H}_T^2 \subset L^2(\Omega, \mathcal{F}_T, \mathbb{P})$, and by the properties of stochastic integrals, it is not difficult to show that $\mathcal{H}_T^2$ is a subspace of $L^2(\Omega, \mathcal{F}_T, \mathbb{P})$. In order to see that $\mathcal{H}_T^2$ is closed, consider a sequence $\{M_T^n\}_{n=1}^{\infty} \subset \mathcal{H}_T^2$, convergent to $M_T \in L^2(\Omega, \mathcal{F}_T, \mathbb{P})$, and write

$$M_T^n = x^n + \int_0^T g_s^n dW_s$$

with $x^n \in \mathbb{R}$, and g^n progressively measurable processes for each $n \in \mathbb{N}$. We only need to show that $M_T \in \mathcal{H}_T^2$.

Itô isometry allows us to write

$$\begin{aligned} \mathbb{E}\left\{\left[(M_T^n - x^n) - (M_T^m - x^m)\right]^2\right\} &= \mathbb{E}\left\{\left[\int_0^T (g_s^n - g_s^m)\, dW_s\right]^2\right\} \\ &= \mathbb{E}\left[\int_0^T (g_s^n - g_s^m)^2\, ds\right] \end{aligned} \tag{2.52}$$

On the other hand, using that $x^n = \mathbb{E}[M_T^n]$, we can write

$$\begin{aligned} \mathbb{E}\left\{\left[(M_T^n - x^n) - (M_T^m - x^m)\right]^2\right\} &= \operatorname{Var}(M_T^n - M_T^m) \\ &\leq \mathbb{E}\left[(M_T^n - M_T^m)^2\right] \end{aligned} \tag{2.53}$$

From (2.52) and (2.53) we obtain

$$\mathbb{E}\left[\int_0^T (g_s^n - g_s^m)^2\, ds\right] \leq \mathbb{E}\left[(M_T^n - M_T^m)^2\right]$$

Since $\{M_T^n\}_{n=1}^{\infty} \to M_T$ in $L^2(\Omega, \mathcal{F}_T, \mathbb{P})$, the right-hand side converges to zero as $m, n \to \infty$. Thus, $\{g^n\}_{n=1}^{\infty}$ is a Cauchy sequence in the space of progressively measurable processes satisfying (2.11), which is a Hilbert space (see Schilling and Partzsch (2012), Lemma 14.18). Then, by completeness, $\{g^n\}_{n=1}^{\infty}$ converges to some process g in the same space.

Let us now prove that $\{M_T^n\}_{n=1}^{\infty}$ converges to

$$\widetilde{M}_T := \mathbb{E}[M_T] + \int_0^T g_s dW_s \tag{2.54}$$

in $L^2(\Omega, \mathcal{F}_T, \mathbb{P})$. On the one hand we have

$$\begin{aligned} \lim_{n\to\infty} \mathbb{E}\left\{\left[\int_0^T g_s^n dW_s - \int_0^T g_s dW_s\right]^2\right\} &= \lim_{n\to\infty} \mathbb{E}\left\{\left[\int_0^T (g_s^n - g_s)\, dW_s\right]^2\right\} \\ &= \lim_{n\to\infty} \mathbb{E}\left[\int_0^T (g_s^n - g_s)^2\, ds\right] = 0 \end{aligned}$$

by the definition of g_s, and therefore,

$$\int_0^T g_s^n dW_s \xrightarrow{L^2} \int_0^T g_s dW_s$$

As L^2-convergence implies L^1-convergence we obtain

$$\lim_{n\to\infty} x^n = \lim_{n\to\infty} \mathbb{E}\left[x^n + \int_0^T g_s^n dW_s\right] = \lim_{n\to\infty} \mathbb{E}\left[M_T^n\right] = \mathbb{E}\left[M_T\right] \tag{2.55}$$

On the other hand

$$\begin{aligned} \|M_T^n - \widetilde{M}_T\|_2 &= \left\|(x^n - \mathbb{E}[M_T]) + \left(\int_0^T g_s^n dW_s - \int_0^T g_s dW_s\right)\right\|_2 \\ &\le |x^n - \mathbb{E}[M_T]| + \left\|\int_0^T g_s^n dW_s - \int_0^T g_s dW_s\right\|_2 \end{aligned}$$

Using (2.55) and Itô isometry, we have that $\lim_{n\to\infty} \|M_T^n - \widetilde{M}_T\|_2 = 0$. By the uniqueness of the limit, $\widetilde{M}_T = M_T$. Since by its definition $\widetilde{M}_T \in \mathcal{H}_T^2$, the proof is complete. □

Theorem 2.85. *Let W be a Brownian motion and $\mathcal{F}_t = \mathcal{F}_t^W$ its natural filtration. Then, every random variable $Y \in L^2(\Omega, \mathcal{F}_T, \mathbb{P})$ can be expressed as*

$$Y = y + \int_0^T g_s dW_s \tag{2.56}$$

with $y = \mathbb{E}[Y]$ and g a progressively measurable process satisfying (2.11). [11]

Proof. To prove the theorem it is enough to show that $L^2(\Omega, \mathcal{F}_T, \mathbb{P}) = \mathcal{H}_T^2$. From Lemma 2.84 we know that $\mathcal{H}_T^2$ is a closed subspace of $L^2(\Omega, \mathcal{F}_T, \mathbb{P})$, and therefore it suffices to prove that $\mathcal{H}_T^2$ is dense in $L^2(\Omega, \mathcal{F}_T, \mathbb{P})$. As $L^2(\Omega, \mathcal{F}_T, \mathbb{P})$ is a Hilbert space, it remains to prove that $\left(\mathcal{H}_T^2\right)^\perp = \{0\}$ (see Theorem A.77).

Let $Y \in \left(\mathcal{H}_T^2\right)^\perp$, then $\mathbb{E}[HY] = 0$ for all $H \in \mathcal{H}_T^2$. Consider a deterministic function $h \in L^2([0, T])$ and define the process

$$X_t = \int_0^t h(s) dW_s$$

with stochastic exponential

$$\mathcal{E}_h(t) := \mathcal{E}(X)_t = \exp\left\{\int_0^t h(s) dW_s - \frac{1}{2}\int_0^t h^2(s) ds\right\}$$

[11] Recall that we are assuming that filtrations satisfy the usual hypotheses, and then $\mathcal{F}_0$ contains all the $\mathbb{P}$-null sets of $\mathcal{F}$.

By Proposition 2.75 we can write

$$\mathcal{E}_h(t) = 1 + \int_0^t \mathcal{E}_h(s)\, dX_s = 1 + \int_0^t \mathcal{E}_h(s)\, h(s)\, dW_s$$

and

$$\mathcal{E}_h(T) = 1 + \int_0^T \mathcal{E}_h(s)\, h(s)\, dW_s$$

Therefore, $\mathcal{E}_h(T) \in \mathcal{H}_T^2$, and $\mathbb{E}[Y\mathcal{E}_h(T)] = 0 \ \ \forall h \in L^2([0,T])$. Since the set

$$\left\{\mathcal{E}_h(T),\ h \in L^2([0,T])\right\}$$

is a total subset of $L^2(\Omega, \mathcal{F}_T, \mathbb{P})$ (see Nualart (2006), Lemma 1.1.2), we obtain $Y = 0$ and $\left(\mathcal{H}_T^2\right)^{\perp} = \{0\}$, which completes the proof.[12] □

Remark 2.86. Theorem 2.85 tells us nothing about the integrand in expression (2.56). It can be obtained, under some conditions, by using Malliavin calculus techniques (see, for example, Nualart (2006)).

Finally, we obtain the desired martingale representation result as a consequence of Theorem 2.85.

Corollary 2.87. ***(Martingale representation)*** *Let M_t be a square-integrable martingale, and let $\mathcal{F}_t$ be the natural filtration of W_t.*[13] *Then, there exists a progressively measurable process g such that*

$$M_t = M_0 + \int_0^t g_s dW_s$$

Proof. By Theorem 2.85

$$M_T = \mathbb{E}[M_T] + \int_0^T g_s dW_s \tag{2.57}$$

with g progressively measurable satisfying (2.11). Taking conditional expectation and applying that $\int_0^t g_s dW_s$ is a martingale, we obtain

$$M_t = \mathbb{E}[M_T \,|\mathcal{F}_t] = \mathbb{E}[\mathbb{E}[M_T]\,|\mathcal{F}_t] + \mathbb{E}\left[\int_0^T g_s dW_s \,|\mathcal{F}_t\right] = M_0 + \int_0^t g_s dW_s$$

which concludes the proof. □

Remark 2.88. It is possible to obtain a martingale representation result valid for the case in which $\mathcal{F}_t = \mathcal{F}_t^{\mathbf{W}}$, where $\mathbf{W} = (W_1, \ldots, W_n)$ is a multidimensional Brownian motion, just by replacing in (2.57) $\int_0^T g_s dW_s$ by $\sum_{i=1}^n \int_0^T g_{is} dW_{is}$ (see Kallianpur and Sundar (2014), Theorem 5.6.2).

[12] A subset A of a Hilbert space H is called total if $\overline{\text{span}(A)} = H$, or equivalently, $A^{\perp} = \{0\}$.

[13] A martingale M_t is a *square-integrable martingale* if $M_t \in L^2(\Omega, \mathcal{F}_t, \mathbb{P})\ \forall t \geq 0$.

2.7 Solutions to the Exercises

Exercise 2.7

The integrand is a progressively measurable step process of the form (2.5), with $k = 1$, $t_i = a$, $t_{i+1} = b$ and $\xi_i(\omega) = 1\ \forall\omega \in \Omega$. Thus, by (2.6) we have

$$\int_0^\infty \mathbf{1}_{(a,b]}(t)dW_t = W_b - W_a$$

Then, if we define

$$\int_a^b dW_t := \int_0^\infty \mathbf{1}_{(a,b]}(t)dW_t$$

we have

$$\int_a^b dW_t = W_b - W_a$$

which is an expected result when defining an integral.

Exercise 2.24

Rewriting the sum on the left hand side of (2.15), we have

$$\sum_{i=1}^{n} \left[(1-\lambda)\ W_{t_{i-1}} + \lambda W_{t_i}\right]\left(W_{t_i} - W_{t_{i-1}}\right)$$

$$= \sum_{i=1}^{n} \left[\lambda W_{t_i}^2 - (1-\lambda)\, W_{t_{i-1}}^2 + (1-2\lambda)\, W_{t_i} W_{t_{i-1}}\right]$$

$$= \frac{1}{2}\sum_{i=1}^{n}\left(W_{t_i}^2 - W_{t_{i-1}}^2\right) + \left(\lambda - \frac{1}{2}\right)\sum_{i=1}^{n}\left(W_{t_i} - W_{t_{i-1}}\right)^2$$

$$= \frac{1}{2}W_t^2 + \left(\lambda - \frac{1}{2}\right)\sum_{i=1}^{n}\left(W_{t_i} - W_{t_{i-1}}\right)^2$$

Therefore, it only remains to show that $\sum_{i=1}^{n}\left(W_{t_i} - W_{t_{i-1}}\right)^2 \to t$ in L^2-sense, but this was already obtained in the proof of Theorem 1.25.

Exercise 2.38

We can write $X_t = f(t, W_t)$, with $f(t,x) = e^{at+bx}$, obtaining

$$\frac{\partial f}{\partial t}(t,x) = af(t,x)\,; \quad \frac{\partial f}{\partial x}(t,x) = bf(t,x)\,; \quad \frac{\partial^2 f}{\partial x^2}(t,x) = b^2 f(t,x)$$

Then, by Itô's formula (2.25)

$$dX_t = \left[aX_t + \frac{1}{2}b^2 X_t\right]dt + bX_t dW_t = \left(a + \frac{b^2}{2}\right)X_t dt + bX_t dW_t$$

Exercise 2.41

By Itô's formula (2.29) we have

$$g(W_t) = g(0) + \frac{1}{2}\int_0^t g''(W_s)\,ds + \int_0^t g'(W_s)\,dW_s$$

and then

$$\begin{aligned}[g(W_\cdot), W_\cdot]_t &= \left[\int_0^\cdot g'(W_s)\,dW_s, W_\cdot\right]_t = \int_0^t g'(W_s)\,d[W_\cdot, W_\cdot]_s \\ &= \int_0^t g'(W_s)\,ds\end{aligned}$$

where we have applied that $\left[\int_0^\cdot g''(W_s)\,ds, W_\cdot\right]_t = 0$, since $\int_0^t g''(W_s)\,ds$ is of finite variation and W_t is continuous.

Exercise 2.42

Taking $f(x) = e^{\alpha x}$, we have $e^{\alpha X_t} = f(X_t)$, and

$$f'(x) = \alpha e^{\alpha x}; \quad f''(x) = \alpha^2 e^{\alpha x}$$

By Itô's formula (2.28) we obtain

$$d\left(e^{\alpha X_t}\right) = \left(\mu\alpha + \frac{1}{2}\sigma^2\alpha^2\right) e^{\alpha X_t}\,dt + \sigma\alpha e^{\alpha X_t}\,dW_t \tag{2.58}$$

which can be expressed as

$$\frac{d\left(e^{\alpha X_t}\right)}{e^{\alpha X_t}} = \left(\mu\alpha + \frac{1}{2}\sigma^2\alpha^2\right) dt + \sigma\alpha dW_t$$

On the other hand, expressing equation (2.58) in integral form, we have

$$e^{\alpha X_t} = 1 + \left(\mu\alpha + \frac{1}{2}\sigma^2\alpha^2\right)\int_0^t e^{\alpha X_s}\,ds + \sigma\alpha\int_0^t e^{\alpha X_s}\,dW_s$$

In order to prove that the stochastic integral is a martingale, we need to check condition (2.11):

$$\begin{aligned}\mathbb{E}\left[\int_0^t e^{2\alpha X_s}\,ds\right] &= \int_0^t \mathbb{E}\left[e^{2\alpha X_s}\right] ds = \int_0^t e^{2\alpha\mu s}\,\mathbb{E}\left[e^{2\alpha\sigma W_s}\right] ds \\ &= \int_0^t e^{2\alpha(\mu+\alpha\sigma^2)s}\,ds < \infty \ \forall t \geq 0\end{aligned}$$

where we have used Tonelli's theorem, that $X_s = \mu s + \sigma W_s$ and Exercise 1.14. Taking expectations and using that the stochastic integral is a martingale, we obtain

$$\mathbb{E}\left[e^{\alpha X_t}\right] = 1 + \left(\mu\alpha + \frac{1}{2}\sigma^2\alpha^2\right)\int_0^t \mathbb{E}\left[e^{\alpha X_s}\right] ds$$

Making $\mathbb{E}\left[e^{\alpha X_t}\right] = f(t)$ and differentiating the previous expression, we obtain the differential equation

$$f'(t) = \left(\mu\alpha + \frac{1}{2}\sigma^2\alpha^2\right) f(t)$$

The solution of this equation with the initial condition $f(0) = \mathbb{E}\left[e^{\alpha X_0}\right] = 1$ is

$$f(t) = \mathbb{E}\left[e^{\alpha X_t}\right] = e^{\left(\mu\alpha + \frac{1}{2}\sigma^2\alpha^2\right)t}$$

Exercise 2.43

Expressing (2.30) in integral form and taking expectations we have

$$\mathbb{E}[X_t] = 1 + \alpha \int_0^t \mathbb{E}[X_s]\, ds$$

Making $f(t) := \mathbb{E}[X_t]$ and differentiating we obtain the differential equation

$$f'(t) = \alpha f(t)$$

whose solution with $f(0) = 1$ is

$$f(t) = \mathbb{E}[X_t] = e^{\alpha t}$$

In order to obtain $d\left(\frac{1}{X_t}\right)$ we consider, for $x \neq 0$, $g(x) = \frac{1}{x}$, from where we get

$$g'(x) = -\frac{1}{x^2}; \quad g''(x) = \frac{2}{x^3}$$

Applying Itô's formula (2.28), we obtain

$$\begin{aligned} d\left(\frac{1}{X_t}\right) &= \left[\alpha X_t \left(-\frac{1}{X_t^2}\right) + \frac{1}{2}\sigma^2 X_t^2 \cdot \frac{2}{X_t^3}\right] dt + \sigma X_t \left(-\frac{1}{X_t^2}\right) dW_t \\ &= \left(-\frac{\alpha}{X_t} + \frac{\sigma^2}{X_t}\right) dt - \frac{\sigma}{X_t} dW_t \end{aligned}$$

which can be expressed in a more compact form, making $Z_t := \frac{1}{X_t}$, as

$$\frac{dZ_t}{Z_t} = \left(-\alpha + \sigma^2\right) dt - \sigma dW_t$$

Exercise 2.46

a) Taking $f(x) = e^x$ we have $\int_0^t e^{W_s} dW_s = \int_0^t f(W_s)\, dW_s$. Applying Corollary 2.45 to the function $F(x) = f(x) = e^x$ we obtain

$$\int_0^t e^{W_s} dW_s = e^{W_s}\Big]_0^t - \frac{1}{2}\int_0^t e^{W_s} ds = e^{W_t} - 1 - \frac{1}{2}\int_0^t e^{W_s} ds$$

b) Let $f(x) = xe^x$. Integrating by parts we obtain

$$F(x) = \int f(x)\,dx = \int xe^x dx = xe^x - e^x + C$$

and

$$\frac{\partial^2 F}{\partial x^2}(x) = \frac{\partial f}{\partial x}(x) = e^x + xe^x$$

Then, by Corollary 2.45 we get

$$\begin{aligned}\int_0^t W_s e^{W_s}\,dW_s &= W_s e^{W_s} - e^{W_s}\Big]_0^t - \frac{1}{2}\int_0^t \left(e^{W_s} + W_s e^{W_s}\right) ds \\ &= (W_t - 1)\,e^{W_t} + 1 - \frac{1}{2}\int_0^t (1 + W_s)\,e^{W_s}\,ds\end{aligned}$$

c) Taking $f(t,x) = e^{t/2}\cos x$ we have

$$F(t,x) = \int f(t,x)\,dx = e^{t/2}\int \cos x dx = e^{t/2}\sin x + C$$

and

$$\frac{\partial^2 F}{\partial x^2}(t,x) = -e^{t/2}\sin x; \quad \frac{\partial F}{\partial t}(t,x) = \frac{1}{2}e^{t/2}\sin x$$

Then, again by Corollary 2.45, we obtain

$$\begin{aligned}\int_0^t e^{s/2}\cos W_s\,dW_s &= e^{s/2}\sin W_s\Big]_0^t - \int_0^t \left(\frac{1}{2}e^{s/2}\sin W_s - \frac{1}{2}e^{s/2}\sin W\right) ds \\ &= e^{t/2}\sin W_t\end{aligned}$$

Exercise 2.56

As f is absolutely continuous, it is differentiable almost everywhere, f' is Lebesgue integrable, and we can write

$$f(t) = f(0) + \int_0^t f'(s)\,ds$$

Thus, $f(t)$ is an Itô process. Applying integration by parts to $f(t)\,W_t$ we get

$$d\left[f(t)\,W_t\right] = f(t)\,dW_t + W_t df(t) + d\left[f(\cdot), W_\cdot\right]_t$$

On the other hand, absolute continuity also implies finite variation, and then, by Theorem 1.64, $[f(\cdot), W_\cdot]_t = 0$. Recalling that $W_0 = 0$, we obtain the desired result.

Exercise 2.57

By integration by parts we can write

$$d\left(\frac{X_t}{Y_t}\right) = d\left(X_t Y_t^{-1}\right) = X_t d\left(Y_t^{-1}\right) + Y_t^{-1} dX_t + d\left[X_\cdot, Y_\cdot^{-1}\right]_t \tag{2.59}$$

To obtain $d\left(Y_t^{-1}\right)$, we apply Itô's formula with $f(x) = x^{-1}$, $x \neq 0$, which gives

$$dY_t^{-1} = \frac{1}{Y_t^3} d\,[Y_\cdot, Y_\cdot]_t - \frac{1}{Y_t^2} dY_t$$

On the other hand, writing this last expression in integral form

$$Y_t^{-1} = Y_0^{-1} + \int_0^t \frac{d\,[Y_\cdot, Y_\cdot]_s}{Y_s^3} - \int_0^t \frac{dY_s}{Y_s^2}$$

we can obtain

$$\left[X_\cdot, Y_\cdot^{-1}\right]_t = -\int_0^t \frac{d\,[X_\cdot, Y_\cdot]_s}{Y_s^2}$$

Replacing the two previous expressions in (2.59) we get

$$\begin{aligned} d\left(\frac{X_t}{Y_t}\right) &= X_t \left(\frac{1}{Y_t^3} d\,[Y_\cdot, Y_\cdot]_t - \frac{1}{Y_t^2} dY_t\right) + Y_t^{-1} dX_t - \frac{1}{Y_t^2} d\,[X_\cdot, Y_\cdot]_t \\ &= Y_t^{-1} dX_t - \frac{X_t}{Y_t^2} dY_t + \frac{X_t}{Y_t^3} d\,[Y_\cdot, Y_\cdot]_t - \frac{1}{Y_t^2} d\,[X_\cdot, Y_\cdot]_t \end{aligned}$$

Exercise 2.78

Since $\int_0^t \operatorname{sgn}^2(W_s)\, ds = \int_0^t ds = t$, we have that

$$\mathbb{P}\left[\int_0^t \operatorname{sgn}^2(W_s)\, ds < \infty\right] = 1$$

and the process X_t is a continuous local martingale. Moreover, $X_0 = 0$, and $[X_\cdot, X_\cdot]_t = \int_0^t \operatorname{sgn}^2(W_s)\, ds = t$, and then, by Lévy's characterization theorem, X_t is a Brownian motion.

2.8 Problems

1. Let f and g be two functions such that $f : [0, \infty) \to \mathbb{R}$ is of class C^1 and $g : \mathbb{R} \to \mathbb{R}$ is of class C^2. Show that

$$\begin{aligned} \int_0^t f(s)\, g'(W_s)\, dW_s = [f(s)\, g(W_s)]_0^t &- \int_0^t f'(s)\, g(W_s)\, ds \\ &- \frac{1}{2}\int_0^t f(s)\, g''(W_s)\, ds \end{aligned}$$

2. Let k be a positive integer. Obtain an expression for $\mathbb{E}\left[W_t^{2k}\right]$ using Itô's formula.

3. Calculate the variance of $\int_0^t e^{W_s}\,dW_s$.

4. Show that for any fixed $T > 0$, $M_t = e^{t/2}\sin W_t$ is a martingale for $0 \le t \le T$.

5. Let Φ denote the distribution function of a standard Gaussian random variable. Prove that, for any fixed $T > 0$, $\Phi\left(\frac{W_t}{\sqrt{T-t}}\right)$ is a martingale for $0 \le t \le T$.

6. Consider the Hermite polynomials $h_n(t,x)$, defined by

$$h_n(t,x) := \frac{(-t)^n}{n!} e^{x^2/2t} \frac{d^n}{dx^n}\left(e^{-x^2/2t}\right) \quad n = 0, 1, \ldots$$

   Prove that, for $n = 0, 1, \ldots$

   a) $h_n(t, W_t)$ is a martingale in $[0, T]$.
   b) $dh_{n+1}(t, W_t) = h_n(t, W_t)\,dW_t$

   (Hint: Use the equality $e^{\lambda x - \lambda^2 t/2} = \sum_{n=0}^{\infty} h_n(t,x)\,\lambda^n$)

7. Let $\mathbf{W}_t = (W_{1t}, \ldots W_{nt})^\top$ be an n-dimensional Brownian motion with $n \ge 2$.

   a) Consider the process $\widetilde{W}_t$ defined by

$$\widetilde{W}_t := \sum_{i=1}^{n} \int_0^t \frac{W_{is}}{\|\mathbf{W}_s\|}\,dW_{is} \tag{2.60}$$

   where $\|\mathbf{W}_s\| = \sqrt{W_{1s}^2 + \cdots + W_{ns}^2}$. Prove that $\widetilde{W}_t$ is a one-dimensional Brownian motion.
   (Hint: (Durrett (1996), page 100) If $\mathbf{W}_t$ is an n-dimensional Brownian motion with $n \ge 2$, starting at $\mathbf{x} \in \mathbb{R}^n$, then

$$\mathbb{P}\{\mathbf{W}_t = 0 \text{ for some } t > 0 \mid \mathbf{W}_0 = \mathbf{x}\} = 0.$$

   Use this fact to justify that the stochastic integral in (2.60) is well defined.)

   b) The squared Bessel process of dimension n is defined by $Z_t := \|\mathbf{W}_t\|^2$. Obtain the stochastic differential of Z_t.

8. Express the following random variables in the form stated in Theorem 2.85.

   a) $Y = W_T$
   b) $Y = \int_0^T W_t\,dt$
   c) $Y = W_T^2$

2.9 Notes

The main references used throughout this chapter have been Karatzas and Shreve (1991), Oksendal (1998), Revuz and Yor (1999), Lipster and Shiryaev (2001), Steele (2001), Klebaner (2005), Protter (2005), Kuo (2006), Mörters and Peres (2010), Schilling and Partzsch (2012), Kallianpur and Sundar (2014), Calin (2015) and Le Gall (2016).

In this chapter we have constructed the stochastic integral with respect to Brownian motion, partially following what was done in Mörters and Peres (2010). This is the most basic notion of stochastic integral required for the purposes of this book. There are other approaches in the literature that consider more general integrators. For example, the stochastic integral is defined with respect to continuous martingales in Bass (2011), right-continuous martingales in Kuo (2006), continuous local martingales in Karatzas and Shreve (1991), continuous semimartingales in Revuz and Yor (1999) and Le Gall (2016), and general semimartingales in Protter (2005).

The Itô integral is used in most applications of stochastic calculus to quantitative finance. The Stratonovich approach can be found, for example, in Bjork and Christensen (1999) and in Bueno-Guerrero et al. (2016). In our presentation of Itô's formula, unlike other textbooks, we have begun by stating the most general result possible within our approach, that is, the formula for functions of time and an Itô process, and then presented the remaining variants as specific cases. We have decided to omit the so-called *box calculus* or *box algebra* (Karatzas and Shreve, 1991; Steele, 2001) involving the multiplication of differentials, since it becomes unnecessary once the rules for calculus with quadratic covariations are known. The part devoted to the generalized Itô formula is somewhat advanced for the general level of the book, but we thought it was convenient to include it to obtain the Tanaka-Meyer formulas. There are different versions of Itô's formula associated with the different concepts of stochastic integral mentioned above, which can be consulted in the corresponding references.

We have presented a version of Bayes' formula for conditional expectations with respect to filtrations, as this is the most commonly used in quantitative finance applications. Versions for general σ-algebras can be found in Björk (2020) and Klebaner (2005).

In this chapter, we have defined the stochastic exponential for Itô processes. A more general definition for semimartingales, along with some properties, can be found in Protter (2005) and Klebaner (2005). Furthermore, the latter reference also studies its inverse, the stochastic logarithm. Our proof of Lévy's Characterization theorem is based on Novikov's condition. Other proofs can be found in Kallianpur and Sundar (2014), who use Itô's formula, or in Le Gall (2016), which is based on the characterization of Brownian motion as a Gaussian process (see Theorem 1.15).

The version of Girsanov's theorem for Brownian motion presented here is the most basic for our purposes. However, for its proof we needed to use results from the theory of stochastic integration with respect to local martingales which are not discussed in this book. Different presentations of Girsanov's theorem can be found in the literature. Thus, for example, a version for multidimensional Brownian motion can be found in Schilling and Partzsch (2012), for continuous martingales in Klebaner (2005), for continuous local martingales in Kallianpur and Sundar (2014), and for semimartingales in Protter (2005).

Regarding the sufficient conditions under which the stochastic exponential of a continuous local martingale is a martingale, we have only presented Novikov's condition, but there are others, such as Kazamaki's criterion. See Protter (2005) for this criterion and a for a more detailed discussion of the subject.

Chapter 3

Stochastic Differential Equations

In this chapter we address the study of stochastic differential equations (SDEs), which describe the time evolution of stochastic processes. This study shares certain analogies with that of ordinary differential equations, analogies which we will explore in the first sections of this chapter.

SDEs are ubiquitous in quantitative finance. Most mathematical models describing the behavior of financial markets incorporate SDEs to describe the evolution of the relevant variables. The relationship between models and SDEs is so close that sometimes models are expressed simply by writing the associated SDE.

The chapter is not limited to the study of SDEs and their integration techniques. It also includes results concerning certain properties of their solutions (Markov and strong Markov properties, diffusive nature) and other results relating SDEs to PDEs (Kolmogorov's equations and the Feynman-Kac formula).

The chapter begins with Section 3.1, which presents the theorem of existence and uniqueness of solutions. Examples are also provided showing how violations of the hypotheses can lead to the non-existence or non-uniqueness of solutions.

Section 3.2 is dedicated to linear SDEs, for which closed-form solutions can be found. We begin by studying the most general case, and then turn to two particular cases of interest: scalar linear SDEs and linear SDEs in the narrow sense.

The results concerning linear SDEs are applied in Section 3.3 to solve three differential equations whose solutions are processes of great importance in the development of stochastic calculus: the mean-reverting Ornstein-Uhlenbeck process, the geometric Brownian motion, and the Brownian bridge. Furthermore, the most relevant properties of these processes are stated.

Section 3.4 presents different integration techniques, analogous to those used with ordinary differential equations: transformation to linear, exact SDEs, and integrating factors. All of these are illustrated with examples.

Section 3.5 explores the Markov and strong Markov properties of the solutions of SDEs, allowing us to apply various concepts related to the theory of Markov processes to these solutions, such as transition probability functions, the Chapman-Kolmogorov equation, transition operators, infinitesimal generators, and Dynkin's formula.

Also related to the theory of Markov processes, Section 3.6 studies diffusion processes, whose importance stems from the fact that many of the processes used in quantitative finance, including Brownian motion, the mean-reverting Ornstein-Uhlenbeck process, and geometric Brownian motion, belong to this class. The main result of this section states that the solutions of SDEs with continuous coefficients are diffusions, and relates the coefficients of the SDE to those of the associated diffusion process.

Section 3.7 is dedicated to two fundamental results in quantitative finance that relate PDEs and SDEs: the Kolmogorov equations, which are two PDEs for the transition probability density of the solution of an SDE, and the Feynman-Kac formula, which provides a probabilistic representation of the solutions to certain PDEs. The chapter concludes with the sections dedicated to problems (Section 3.9) and notes (Section 3.10).

3.1 Existence and Uniqueness of Solutions

Let $\mathbf{b} : [0,T] \times \mathbb{R}^n \to \mathbb{R}^n$ and $\mathbf{\Sigma} : [0,T] \times \mathbb{R}^n \to \mathbb{R}^{n \times d}$ be measurable functions, with

$$\mathbf{b} = (b_1, \ldots, b_n)^\top, \quad \mathbf{\Sigma} = \begin{pmatrix} \sigma_{11} & \cdots & \sigma_{1d} \\ \vdots & \ddots & \vdots \\ \sigma_{n1} & \cdots & \sigma_{nd} \end{pmatrix}$$

and Euclidean norms

$$||\mathbf{b}||^2 = \sum_{i=1}^{n} b_i^2, \quad ||\mathbf{\Sigma}||^2 = \sum_{i=1}^{n} \sum_{j=1}^{d} \sigma_{ij}^2$$

For a process to be considered solution of a stochastic differential equation (SDE), it must satisfy certain conditions, which appear below.

Definition 3.1. (Solution of an SDE) Let $\mathbf{W}_t$ be a d-dimensional Brownian motion. An n-dimensional process $\mathbf{X}_t$ is a solution of the SDE

$$d\mathbf{X}_t = \mathbf{b}(t, \mathbf{X}_t)\, dt + \mathbf{\Sigma}(t, \mathbf{X}_t)\, d\mathbf{W}_t, \quad 0 \le t \le T, \quad \mathbf{X}_0 = \xi \tag{3.1}$$

if it satisfies:

i) $\mathbf{X}_t$ has continuous sample paths and is adapted to the filtration $\mathcal{F}_t^{\xi,\mathbf{W}}$ generated by ξ and $\mathbf{W}$.

ii) $\int_0^T \left(||\mathbf{b}(t, \mathbf{X}_t)|| + ||\mathbf{\Sigma}(t, \mathbf{X}_t)||^2 \right) dt < \infty$ a.s.

iii) For each $t \in [0,T]$

$$\mathbf{X}_t = \xi + \int_0^t \mathbf{b}(s, \mathbf{X}_s)\, ds + \int_0^t \mathbf{\Sigma}(s, \mathbf{X}_s)\, d\mathbf{W}_s \quad \text{a.s.}$$

Remark 3.2. Condition i) in Definition 3.1 implies that the solution is a progressively measurable process. Condition ii) ensures that the integrals in iii) are well-defined. Finally, by condition iii), the process $\mathbf{X}_t$ satisfies the integral form of (3.1) for every $t \in [0,T]$, and is, therefore, a global solution.

Remark 3.3. The concept of solution presented here is also known as *strong solution*, to differentiate it from another type of solution, that requires weaker conditions, called *weak solution*, which we will not address here (see, for example, Karatzas and Shreve (1991) or Oksendal (1998)).

In ordinary calculus, the existence and the uniqueness of solutions of ordinary differential equations (ODEs) are obtained imposing Lipschitz and linear growth conditions. A similar situation occurs with stochastic differential equations.

Theorem 3.4. ***(Existence and uniqueness)*** *Let $\boldsymbol{\xi}$ be a random variable independent of $\sigma(\mathbf{W}_s, 0 \le s \le T)$, and such that $\mathbb{E}\left[\|\boldsymbol{\xi}\|^2\right] < \infty$.*[1] *Assume that the functions* $\mathbf{b}$ *and* $\boldsymbol{\Sigma}$ *satisfy, for some constants C and D, the Lipschitz condition*

$$\|\mathbf{b}(t,\mathbf{x}) - \mathbf{b}(t,\mathbf{y})\| + \|\boldsymbol{\Sigma}(t,\mathbf{x}) - \boldsymbol{\Sigma}(t,\mathbf{y})\| \le C\|\mathbf{x}-\mathbf{y}\| \quad \forall \mathbf{x},\mathbf{y} \in \mathbb{R}^n, \quad \forall t \in [0,T] \tag{3.2}$$

and the linear growth condition

$$\|\mathbf{b}(t,\mathbf{x})\| + \|\boldsymbol{\Sigma}(t,\mathbf{x})\| \le D(1+\|\mathbf{x}\|) \quad \forall \mathbf{x} \in \mathbb{R}^n, \quad \forall t \in [0,T] \tag{3.3}$$

Then, the stochastic differential equation

$$d\mathbf{X}_t = \mathbf{b}(t,\mathbf{X}_t)\,dt + \boldsymbol{\Sigma}(t,\mathbf{X}_t)\,d\mathbf{W}_t, \quad 0 \le t \le T, \quad \mathbf{X}_0 = \boldsymbol{\xi}$$

has a unique solution in the sense of Definition 3.1 satisfying $\mathbb{E}\left[\int_0^T \|\mathbf{X}_t\|^2\,dt\right] < \infty$.

Proof. See Oksendal (1998), Theorem 5.2.1 □

Remark 3.5. The uniqueness of the solutions must be understood in the following sense. If $\mathbf{X}_{1t}$ and $\mathbf{X}_{2t}$ are two of the solutions whose existence is asserted by Theorem 3.4, then $\mathbb{P}\{\|\mathbf{X}_{1t} - \mathbf{X}_{2t}\| = 0 \text{ for all } t \in [0,T]\} = 1$.

Remark 3.6. In the proof of Theorem 3.4, the Lipschitz condition is used to obtain uniqueness, whereas the linear growth condition allows to obtain the global existence of a solution by using Picard iteration, similar to what is done with ordinary differential equations.

Remark 3.7. Under the conditions of existence and uniqueness of the solution, it is possible to show that there exists a function F such that the solution is given by $\mathbf{X}_t = F\left(t, \{\mathbf{W}_s\}_{0\le s\le t}\right)$, although the function F is, in general, unknown (see Klebaner (2005), Remark 5.2 and references therein).

Remark 3.8. In the one-dimensional case, it is possible to relax the conditions in Theorem 3.4 and still obtain existence and uniqueness (see, for example, Theorem 5.5 in Klebaner (2005) or Proposition 2.13 and Remark 2.16 in Chapter 5 of Karatzas and Shreve (1991)).

The following exercises, taken from Kuo (2006), show how the properties of existence of a global solution and its uniqueness can be lost if the Lipschitz or the linear growth conditions are not satisfied.

Exercise 3.9. Consider the following one-dimensional SDE

$$dX_t = 3X_t^{1/3}dt + 3X_t^{2/3}dW_t, \quad X_0 = 0 \tag{3.4}$$

a) Show that the SDE (3.4) does not satisfy the Lipschitz condition (3.2).

b) Prove that the process $X_t^{(a)} = (W_t - a)^3\,\mathbf{1}_{\{W_t \ge a\}}$ is a solution of (3.4) for any $a \in \mathbb{R}^+$.

[1] In what follows, we will assume that the initial value $\mathbf{X}_0 = \boldsymbol{\xi}$ always satisfy these conditions.

Exercise 3.10. Consider the following one-dimensional SDE

$$dX_t = X_t^3 dt + X_t^2 dW_t, \quad X_0 = 1 \tag{3.5}$$

a) Show that the SDE (3.5) does not satisfy the linear growth condition (3.3).

b) Prove that $X_t = \dfrac{1}{1 - W_t}$ is a solution of (3.5) in $[0, T_1)$, where T_1 is the hitting time of W_t to the level 1: $T_1(\omega) = \inf\{t > 0 : W_t(\omega) = 1\}$.

3.2 Linear SDEs

In the vast majority of cases, it is not possible to find an explicit solution of an SDE, in the sense of Remark 3.7, and numerical methods must be used. However, in the case of linear SDEs, it is possible to obtain closed-form solutions.

Definition 3.11. (Linear SDE) We say that the SDE

$$d\mathbf{X}_t = \mathbf{b}(t, \mathbf{X}_t)\, dt + \mathbf{\Sigma}(t, \mathbf{X}_t)\, d\mathbf{W}_t, \; 0 \le t \le T, \; \mathbf{X}_0 = \xi$$

is a linear SDE if its coefficients are of the form

$$\mathbf{b}(t, \mathbf{X}_t) = \mathbf{F}(t)\, \mathbf{X}_t + \mathbf{f}(t) \tag{3.6}$$

$$\mathbf{\Sigma}(t, \mathbf{X}_t) = \mathbf{G}(t)\, \mathbf{X}_t + \mathbf{g}(t) \tag{3.7}$$

where $\mathbf{f} : [0,T] \to \mathbb{R}^n$, $\mathbf{g} : [0,T] \to \mathbb{R}^{n\times d}$, $\mathbf{F} : [0,T] \to \mathbb{R}^{n\times n}$, $\mathbf{G} : [0,T] \to L\left(\mathbb{R}^n, \mathbb{R}^{n\times d}\right)$, and $L\left(\mathbb{R}^n, \mathbb{R}^{n\times d}\right)$ is the space of bounded linear mappings from $\mathbb{R}^n$ to $\mathbb{R}^{n\times d}$.

Remark 3.12. Using that $\mathbf{f}, \mathbf{g}, \mathbf{F}, \mathbf{G}$ are bounded, and the properties of the norms, it is not difficult to show that $\mathbf{b}(t, \mathbf{x})$ and $\mathbf{\Sigma}(t, \mathbf{x})$ satisfy the Lipschitz and linear growth conditions (3.2) and (3.3). Thus, assuming the conditions on the initial value $\mathbf{X}_0 = \xi$, Theorem 3.4 guarantees the existence and uniqueness of solutions of linear SDEs.

In order to obtain the solution of the general linear SDE, it is useful to write it in the equivalent form

$$d\mathbf{X}_t = [\mathbf{F}(t)\, \mathbf{X}_t + \mathbf{f}(t)]\, dt + \sum_{k=1}^{d} [\mathbf{G}_k(t)\, \mathbf{X}_t + \mathbf{g}_k(t)]\, d\mathbf{W}_t^{(k)} \tag{3.8}$$

where $\mathbf{G}_k : [0,T] \to \mathbb{R}^{n\times n}$, $\mathbf{g}_k : [0,T] \to \mathbb{R}^n$ and $\mathbf{W}_t^{(k)}$ is the k-th component of the d-dimensional Brownian motion $\mathbf{W}_t$. The linear SDE (3.8) is said to be *homogeneous* if $\mathbf{f} = \mathbf{g}_1 = \cdots \mathbf{g}_d = \mathbf{0}$ and it is *linear in the narrow sense* if $\mathbf{G}_1 = \cdots = \mathbf{G}_d = \mathbf{0}$.

To obtain the solution of the complete linear SDE (3.8), we begin by studying the corresponding homogeneous equation, in a manner analogous to what is done in the study of linear ODEs.

Consider the homogeneous linear SDE

$$d\mathbf{X}_t = \mathbf{F}(t)\, \mathbf{X}_t dt + \sum_{k=1}^{d} \mathbf{G}_k(t)\, \mathbf{X}_t d\mathbf{W}_t^{(k)} \tag{3.9}$$

Let $\mathbf{e}_j$ be the j-th vector of the canonical orthonormal basis of $\mathbb{R}^n$, i.e. $\mathbf{e}_j = (0, \cdots, 0, \overset{j)}{1}, 0, \cdots, 0)^\top$, and let $\mathbf{\Phi}_j(t) := (\Phi_{1j}(t), \ldots, \Phi_{nj}(t))^\top$ be the solution of equation (3.9) with initial value $\mathbf{X}_0 = \mathbf{e}_j$.

Definition 3.13. (Fundamental matrix) The $n \times n$ matrix $\mathbf{\Phi}(t)$ defined by

$$\mathbf{\Phi}(t) := (\mathbf{\Phi}_1(t) | \cdots | \mathbf{\Phi}_n(t)) = (\Phi_{ij}(t))_{n \times n}$$

is called the fundamental matrix of the homogeneous linear SDE (3.9).

From the definition of fundamental matrix we have that $\mathbf{\Phi}(0) = \mathbf{I}_n$, where $\mathbf{I}_n$ is the identity matrix of order n, and that $\mathbf{\Phi}(t)$ satisfies the SDE

$$d\mathbf{\Phi}(t) = \mathbf{F}(t)\mathbf{\Phi}(t)\,dt + \sum_{k=1}^{d} \mathbf{G}_k(t)\mathbf{\Phi}(t)\,d\mathbf{W}_t^{(k)} \tag{3.10}$$

which can be expressed componentwise in the form

$$d\Phi_{ij}(t) = \sum_{l=1}^{n} F_{il}(t)\Phi_{lj}(t)\,dt + \sum_{k=1}^{d}\sum_{l=1}^{n} (\mathbf{G}_k)_{il}(t)\Phi_{lj}(t)\,d\mathbf{W}_t^{(k)}$$

Proposition 3.14. *The unique solution of the homogeneous linear SDE (3.9), with initial value* $\mathbf{X}_0 = \xi$, *is given by*

$$\mathbf{X}_t = \mathbf{\Phi}(t)\,\xi \tag{3.11}$$

Proof. We only need to show that (3.11) satisfies equation (3.9).

$$\begin{aligned} d\mathbf{X}_t &= d\mathbf{\Phi}(t)\,\xi = \mathbf{F}(t)\mathbf{\Phi}(t)\,\xi dt + \sum_{k=1}^{d} \mathbf{G}_k(t)\mathbf{\Phi}(t)\,\xi d\mathbf{W}_t^{(k)} \\ &= \mathbf{F}(t)\mathbf{X}_t dt + \sum_{k=1}^{d} \mathbf{G}_k(t)\mathbf{X}_t d\mathbf{W}_t^{(k)} \end{aligned}$$

□

Definition 3.15. (Stochastic Wronskian determinant) The determinant of the fundamental matrix $\mathbf{\Phi}(t)$, $\mathcal{W}(t)$, is called the stochastic Wronskian determinant

$$\mathcal{W}(t) := \det[\mathbf{\Phi}(t)]$$

Note that, since $\mathbf{\Phi}(0) = \mathbf{I}_n$, we have that $\mathcal{W}(0) = 1$.

Theorem 3.16. ***(Stochastic Liouville formula)*** *The stochastic Wronskian determinant associated to equation (3.9) is given by*

$$\mathcal{W}(t) = \exp\left\{\int_0^t \left[\operatorname{tr}(\mathbf{F}(s)) - \frac{1}{2}\sum_{k=1}^{d} \operatorname{tr}\left(\mathbf{G}_k^2(s)\right)\right] ds + \sum_{k=1}^{d}\int_0^t \operatorname{tr}(\mathbf{G}_k(s))\,dW_s^{(k)}\right\}$$

where tr *is the trace of a square matrix.*

Proof. See Mao (2008), page 94. □

Corollary 3.17. *The fundamental matrix* $\mathbf{\Phi}(t)$ *is a.s. invertible for every* $t \in [0, T]$.

Proof. From Theorem 3.16 we have that $\det[\mathbf{\Phi}(t)] = \mathcal{W}(t) > 0$ a.s. □

Now we can state the result with the solution of the general linear SDE.

Theorem 3.18. *The unique solution of the SDE (3.8) is given by*

$$\mathbf{X}_t = \mathbf{\Phi}(t)\left\{\mathbf{X}_0 + \int_0^t \mathbf{\Phi}^{-1}(s)\left[\mathbf{f}(s) - \sum_{k=1}^d \mathbf{G}_k(s)\,\mathbf{g}_k(s)\right]ds \right.$$
$$\left. + \sum_{k=1}^d \int_0^t \mathbf{\Phi}^{-1}(s)\,\mathbf{g}_k(s)\,dW_s^{(k)}\right\}$$

where $\mathbf{\Phi}(t)$ *is the fundamental matrix of the homogeneous SDE (3.9).*

Proof. Consider the process

$$\varphi_t = \mathbf{X}_0 + \int_0^t \mathbf{\Phi}^{-1}(s)\left[\mathbf{f}(s) - \sum_{k=1}^d \mathbf{G}_k(s)\,\mathbf{g}_k(s)\right]ds$$
$$+ \sum_{k=1}^d \int_0^t \mathbf{\Phi}^{-1}(s)\,\mathbf{g}_k(s)\,dW_s^{(k)}$$

which in differential form is given by

$$d\varphi_t = \mathbf{\Phi}^{-1}(t)\left[\mathbf{f}(t) - \sum_{k=1}^d \mathbf{G}_k(t)\,\mathbf{g}_k(t)\right]dt$$
$$+ \sum_{k=1}^d \mathbf{\Phi}^{-1}(t)\,\mathbf{g}_k(t)\,dW_t^{(k)}$$

with $\varphi_0 = \mathbf{X}_0$. Now define the new process $\boldsymbol{\eta}_t := \mathbf{\Phi}(t)\,\varphi_t$. By integration by parts we have

$$\begin{aligned}
d\boldsymbol{\eta}_t &= \varphi_t d\mathbf{\Phi}(t) + \mathbf{\Phi}(t)\,d\varphi_t + d\left[\mathbf{\Phi}(\cdot), \varphi_\cdot\right]_t \\
&= \mathbf{F}(t)\,\boldsymbol{\eta}_t dt + \sum_{k=1}^d \mathbf{G}_k(t)\,\boldsymbol{\eta}_t dW_t^{(k)} \\
&\quad + \left[\mathbf{f}(t) - \sum_{k=1}^d \mathbf{G}_k(t)\,\mathbf{g}_k(t)\right]dt + \sum_{k=1}^d \mathbf{g}_k(t)\,dW_t^{(k)} \\
&\quad + \sum_{k=1}^d \mathbf{G}_k(t)\,\mathbf{g}_k(t)\,dt \\
&= \left[\mathbf{F}(t)\,\boldsymbol{\eta}_t + \mathbf{f}(t)\right]dt + \sum_{k=1}^d \left[\mathbf{G}_k(t)\,\boldsymbol{\eta}_t + \mathbf{g}_k(t)\right]dW_t^{(k)}
\end{aligned}$$

Moreover, $\boldsymbol{\eta}_0 := \boldsymbol{\Phi}(0)\,\boldsymbol{\varphi}_0 = \mathbf{I}_n\boldsymbol{\varphi}_0 = \boldsymbol{\varphi}_0 = \mathbf{X}_0$. Then, we have proved that $\boldsymbol{\eta}_t$ is a solution of equation (3.8), and the proof is concluded. □

In what follows, we will study several particular cases of the general linear SDE (3.8).

Corollary 3.19. *(Scalar linear SDE) Let $F, f, G_k, g_k : [0,T] \to \mathbb{R}$ be measurable and bounded functions. Then, the unique solution of the SDE*

$$dX_t = [F(t)\,X_t + f(t)]\,dt + \sum_{k=1}^{d} [G_k(t)\,X_t + g_k(t)]\,dW_t^{(k)} \tag{3.12}$$

is given by

$$X_t = \Phi(t)\left\{X_0 + \int_0^t \Phi^{-1}(s)\left[f(s) - \sum_{k=1}^{d} G_k(s)\,g_k(s)\right]ds + \sum_{k=1}^{d}\int_0^t \Phi^{-1}(s)\,g_k(s)\,dW_s^{(k)}\right\} \tag{3.13}$$

where

$$\Phi(t) = \exp\left\{\int_0^t \left[F(s) - \frac{1}{2}\sum_{k=1}^{d} G_k^2(s)\right]ds + \sum_{k=1}^{d}\int_0^t G_k(s)\,dW_s^{(k)}\right\} \tag{3.14}$$

Proof. By expression (3.10), the fundamental solution associated to equation (3.12) satisfies the SDE

$$\begin{aligned} d\Phi(t) &= F(t)\,\Phi(t)\,dt + \sum_{k=1}^{d} G_k(t)\,\Phi(t)\,dW_t^{(k)} \\ &= \Phi(t)\left[F(t)\,dt + \sum_{k=1}^{d} G_k(t)\,dW_t^{(k)}\right] \\ &:= \Phi(t)\,dX_t \end{aligned}$$

with $\Phi(0) = 1$. The solution of this SDE is given by

$$\Phi(t) = \mathcal{E}(X)_t = \exp\left\{\int_0^t \left[F(s) - \frac{1}{2}\sum_{k=1}^{d} G_k^2(s)\right]ds + \sum_{k=1}^{d}\int_0^t G_k(s)\,dW_s^{(k)}\right\}$$

and taking into account Theorem 3.18 for the case $n = 1$, the proof is completed. □

Exercise 3.20. Obtain the solution of the SDE

$$dX_t = \sigma X_t\,dW_t$$

Corollary 3.21. ***(Linear SDE in the narrow sense)*** *Let* $\mathbf{F}$, $\mathbf{f}$ *and* $\mathbf{g}_k$ *be as in equation (3.8). Then, the unique solution of the SDE*

$$d\mathbf{X}_t = [\mathbf{F}(t)\mathbf{X}_t + \mathbf{f}(t)]\,dt + \sum_{k=1}^{d} \mathbf{g}_k(t)\,d\mathbf{W}_t^{(k)} \tag{3.15}$$

is given by

$$\mathbf{X}_t = \mathbf{\Phi}(t)\left[\mathbf{X}_0 + \int_0^t \mathbf{\Phi}^{-1}(s)\,\mathbf{f}(s)\,ds + \sum_{k=1}^{d}\int_0^t \mathbf{\Phi}^{-1}(s)\,\mathbf{g}_k(s)\,d\mathbf{W}_s^{(k)}\right] \tag{3.16}$$

where $\mathbf{\Phi}(t)$ *is the solution of the ODE*

$$\frac{d\mathbf{\Phi}(t)}{dt} = \mathbf{F}(t)\,\mathbf{\Phi}(t), \quad \mathbf{\Phi}(0) = \mathbf{I}_n$$

Moreover, if $\mathbf{X}_0$ *is Gaussian, then* $\mathbf{X}_t$ *is a Gaussian process.* [2]

Proof. The proof of the solution is immediate from Theorem 3.18 and expression (3.10), with $\mathbf{G}_1 = \cdots = \mathbf{G}_d = \mathbf{0}$. The Gaussian property is a direct consequence of expression (3.16), considering that, in this case, $\mathbf{\Phi}(t)$ is deterministic as the solution of an ODE. □

3.3 Examples

In this section we carry out a detailed study of several linear SDEs that play a central role in the applications of stochastic calculus, especially in quantitative finance.

Mean-Reverting Ornstein-Uhlenbeck Process

The *mean-reverting Ornstein-Uhlenbeck (OU) process* is determined by the SDE

$$dX_t = -\alpha(X_t - \mu)\,dt + \sigma dW_t \tag{3.17}$$

where α, μ and σ are constants, with $\alpha, \sigma > 0$.

The most important properties of the process are summarized in the following result.

Proposition 3.22. ***(Properties of the mean-reverting OU process)*** *The unique solution of equation (3.17) is*

$$X_t = e^{-\alpha t}X_0 + \mu\left(1 - e^{-\alpha t}\right) + \sigma\int_0^t e^{-\alpha(t-s)}\,dW_s \tag{3.18}$$

and satisfies:

i) $\mathbb{E}[X_t] = e^{-\alpha t}\mathbb{E}[X_0] + \mu\left(1 - e^{-\alpha t}\right)$

[2] See Karatzas and Shreve (1991), problem 6.1 of Chapter 5, for the explicit expressions of the mean vector and the covariance matrix function.

ii) $\text{var}[X_t] = e^{-2\alpha t}\,\text{var}[X_0] + \frac{\sigma^2}{2\alpha}\left(1 - e^{-2\alpha t}\right)$

iii) $\lim_{t\to\infty} \mathbb{E}[X_t] = \mu$

iv) $\lim_{t\to\infty} \text{var}[X_t] = \frac{\sigma^2}{2\alpha}$

v) *X_t converges in distribution to* $N\left(\mu, \frac{\sigma^2}{2\alpha}\right)$

vi) *If X_0 is normally distributed, then X_t is a Gaussian process with mean and covariance functions given by*

$$\mu(t) = e^{-\alpha t}\mathbb{E}[X_0] + \mu\left(1 - e^{-\alpha t}\right)$$

$$c(s,t) = e^{-\alpha(s+t)}\,\text{var}[X_0] + \frac{\sigma^2}{2\alpha}\left[e^{-\alpha|s-t|} - e^{-\alpha(s+t)}\right]$$

vii) *If* $X_0 \sim N\left(\mu, \frac{\sigma^2}{2\alpha}\right)$, *then* $X_t \sim N\left(\mu, \frac{\sigma^2}{2\alpha}\right)$ *for any* $t \geq 0$.

Proof. Equation (3.17) is the linear SDE in the narrow sense (3.15) with $d = 1$, $\mathbf{F}(t) = -\alpha$, $\mathbf{f}(t) = \alpha\mu$ and $\mathbf{g}_1(t) = \sigma$. Solving the ODE

$$\frac{d\Phi(t)}{dt} = -\alpha\Phi(t), \quad \Phi(0) = 1$$

we obtain $\Phi(t) = e^{-\alpha t}$, and then, by (3.16) the solution is

$$\begin{aligned} X_t &= e^{-\alpha t}\left[X_0 + \alpha\mu \int_0^t e^{\alpha s}ds + \sigma \int_0^t e^{\alpha s}dW_s\right] \\ &= e^{-\alpha t}X_0 + \mu\left(1 - e^{-\alpha t}\right) + \sigma \int_0^t e^{-\alpha(t-s)}dW_s \end{aligned}$$

i) It is immediate from expression (3.18), taking into account that stochastic integrals have zero mean.

ii) From (3.18) we have

$$\begin{aligned} \text{var}[X_t] &= e^{-2\alpha t}\,\text{var}[X_0] + \sigma^2\text{var}\left[\int_0^t e^{-\alpha(t-s)}dW_s\right] \\ &= e^{-2\alpha t}\,\text{var}[X_0] + \sigma^2 \int_0^t e^{-2\alpha(t-s)}ds \\ &= e^{-2\alpha t}\,\text{var}[X_0] + \frac{\sigma^2}{2\alpha}\left(1 - e^{-2\alpha t}\right) \end{aligned}$$

where we have applied that X_0 and W_t are independent and Itô isometry.

iii) and iv) are immediate from i) and ii).

v) The characteristic function of X_t is

$$\begin{aligned}\varphi_t(u) &= \mathbb{E}\left[\exp\left(iuX_t\right)\right] \\ &= \exp\left[iu\mu\left(1-e^{-\alpha t}\right)\right]\mathbb{E}\left[\exp\left(iue^{-\alpha t}X_0\right)\right] \\ &\times\exp\left(-\frac{\sigma^2u^2}{4\alpha}\left(1-e^{-2\alpha t}\right)\right)\end{aligned} \tag{3.19}$$

where we have used that, by viii) of Proposition 2.13, $\int_0^t e^{-\alpha(t-s)}dW_s$ has a centered Gaussian distribution with variance $\frac{1}{2\alpha}\left(1-e^{-2\alpha t}\right)$. Now, as $\left|\exp\left(iue^{-\alpha t}X_0\right)\right| \le 1$, by the dominated convergence theorem we have

$$\lim_{t\to\infty}\mathbb{E}\left[\exp\left(iue^{-\alpha t}X_0\right)\right] = \mathbb{E}\left[\lim_{t\to\infty}\exp\left(iue^{-\alpha t}X_0\right)\right] = 1$$

and therefore

$$\lim_{t\to\infty}\varphi_t(u) = \exp\left(iu\mu\right)\exp\left(-\frac{\sigma^2u^2}{4\alpha}\right) = \exp\left[iu\mu - \frac{1}{2}\left(\frac{\sigma^2}{2\alpha}\right)u^2\right]$$

Then, by Lévy's continuity theorem, X_t converges in distribution to a $N\left(\mu, \frac{\sigma^2}{2\alpha}\right)$ random variable.

vi) X_t is a Gaussian process by Corollary 3.21. For the mean and covariance functions, we have

$$\mu(t) = \mathbb{E}\left[X_t\right] = e^{-\alpha t}\mathbb{E}\left[X_0\right] + \mu\left(1-e^{-\alpha t}\right)$$

$$\begin{aligned}c(s,t) &= \operatorname{cov}\left(X_s, X_t\right) \\ &= e^{-\alpha(s+t)}\operatorname{var}\left[X_0\right] + \operatorname{cov}\left[\sigma\int_0^s e^{-\alpha(s-u)}dW_u, \sigma\int_0^t e^{-\alpha(t-u)}dW_u\right] \\ &= e^{-\alpha(s+t)}\operatorname{var}\left[X_0\right] + \sigma^2\int_0^{s\wedge t} e^{-\alpha(s-u)}e^{-\alpha(t-u)}du \\ &= e^{-\alpha(s+t)}\operatorname{var}\left[X_0\right] + \sigma^2 e^{-\alpha(s+t)}\int_0^{s\wedge t} e^{2\alpha u}du \\ &= e^{-\alpha(s+t)}\operatorname{var}\left[X_0\right] + \frac{\sigma^2}{2\alpha}e^{-\alpha(s+t)}\left(e^{2\alpha(s\wedge t)}-1\right) \\ &= e^{-\alpha(s+t)}\operatorname{var}\left[X_0\right] + \frac{\sigma^2}{2\alpha}\left[e^{-\alpha|s-t|} - e^{-\alpha(s+t)}\right]\end{aligned}$$

vii) If $X_0 \sim N\left(\mu, \frac{\sigma^2}{2\alpha}\right)$, then $e^{-\alpha t}X_0 \sim N\left(e^{-\alpha t}\mu, e^{-2\alpha t}\frac{\sigma^2}{2\alpha}\right)$. Thus, from expression (3.19), we have

$$\varphi_t(u) = \exp\left[iu\mu\left(1 - e^{-\alpha t}\right)\right]\exp\left(iue^{-\alpha t}\mu - \frac{1}{2}e^{-2\alpha t}\frac{\sigma^2}{2\alpha}u^2\right)$$
$$\times \exp\left[-\frac{\sigma^2 u^2}{4\alpha}\left(1 - e^{-2\alpha t}\right)\right]$$
$$= \exp\left[iu\mu - \frac{1}{2}\left(\frac{\sigma^2}{2\alpha}\right)u^2\right]$$

which is the characteristic function of a $N\left(\mu, \frac{\sigma^2}{2\alpha}\right)$ random variable.

□

The term "mean-reverting" refers to the fact that, according to iii) of Proposition 3.22, as $t \to \infty$, the mean of the process approaches μ, which is known as *long-term mean* of the process. The rate at which $\mathbb{E}[X_t]$ approaches μ is governed by the parameter α, which is called *mean-reversion speed*. The higher the value of α, the faster the approach to the long-term mean.

The mean-reverting OU process is widely used in quantitative finance, especially in interest rate modeling. Figure 3.1 shows one sample path of a mean-reverting OU process started at $X_0 = 0.2$, with $\alpha = 0.8$, $\mu = 0.05$ and $\sigma = 0.05$.

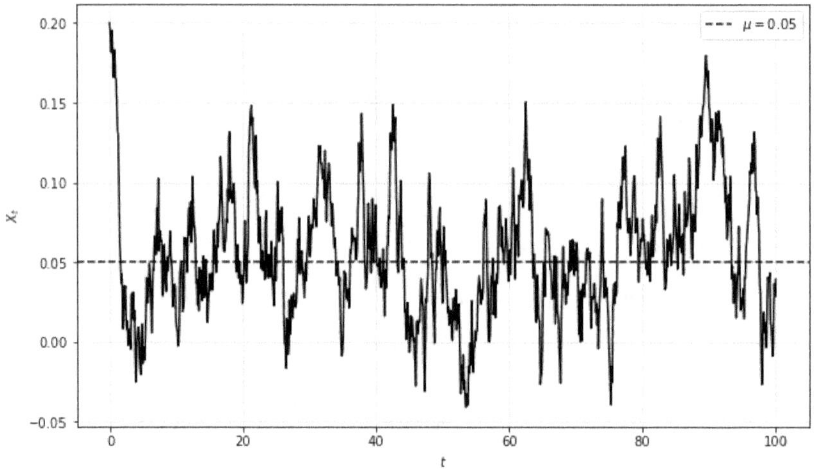

Figure 3.1: A mean-reverting OU process path.

A particular case of interest of equation (3.17) occurs when $\mu = 0$, since we obtain the Langevin equation (2.2) that we studied as an example in Section 2.1. The solution of this equation is known as *Ornstein-Uhlenbeck (OU) process* and is given by equation (2.3). This solution, and the other properties of the OU process, can be obtained by simply setting $\mu = 0$ in Proposition 3.22.

Geometric Brownian Motion

The geometric Brownian motion (GBM) is the solution of the SDE

$$dX_t = \mu X_t dt + \sigma X_t dW_t \tag{3.20}$$

with μ and σ two constants.

Proposition 3.23. ***(Properties of GBM)*** *The unique solution of equation (3.20) is*

$$X_t = X_0 \exp\left[\left(\mu - \frac{\sigma^2}{2}\right)t + \sigma W_t\right] \tag{3.21}$$

and satisfies:

i) $\mathbb{E}[X_t] = \mathbb{E}[X_0] e^{\mu t}$

ii) $\text{var}[X_t] = \mathbb{E}\left[X_0^2\right] \exp\left[2\left(\mu + \frac{\sigma^2}{2}\right)t\right] - (\mathbb{E}[X_0])^2 \exp(2\mu t)$

iii) X_t *is a martingale if, and only if,* $\mu = 0$.

iv) If $X_0 > 0$ *a.s. and* $\ln X_0$ *is normally distributed, then* $Y_t := \ln X_t$ *is a Gaussian process with mean and covariance functions given by*

$$\mu(t) = \mathbb{E}[Y_0] + \left(\mu - \frac{\sigma^2}{2}\right)t$$

$$c(s,t) = \text{var}[Y_0] + \sigma^2 (s \wedge t)$$

Proof. Equation (3.20) is a scalar linear SDE, determined by expression (3.12), with $d = 1$, $f = g_k = 0$, $F(t) = \mu$ and $G_1(t) = \sigma$. Then

$$\Phi(t) = \exp\left[\left(\mu - \frac{\sigma^2}{2}\right)t + \sigma W_t\right]$$

and

$$X_t = \Phi(t) X_0 = X_0 \exp\left[\left(\mu - \frac{\sigma^2}{2}\right)t + \sigma W_t\right]$$

i) From (3.21) we have

$$\begin{aligned}\mathbb{E}[X_t] &= \mathbb{E}[X_0] \exp\left[\left(\mu - \frac{\sigma^2}{2}\right)t\right] \mathbb{E}[\exp(\sigma W_t)] \\ &= \mathbb{E}[X_0] \exp\left[\left(\mu - \frac{\sigma^2}{2}\right)t\right] \exp\left[\frac{\sigma^2}{2}t\right] \\ &= \mathbb{E}[X_0] e^{\mu t}\end{aligned}$$

ii) Using again (3.21) we can write

$$\begin{aligned}\mathbb{E}\left[X_t^2\right] &= \mathbb{E}\left\{X_0^2 \exp\left[2\left(\mu - \frac{\sigma^2}{2}\right)t + 2\sigma W_t\right]\right\} \\ &= \mathbb{E}\left[X_0^2\right] \exp\left[2\left(\mu - \frac{\sigma^2}{2}\right)t\right] \mathbb{E}\left[\exp\left(2\sigma W_t\right)\right] \\ &= \mathbb{E}\left[X_0^2\right] \exp\left[2\left(\mu - \frac{\sigma^2}{2}\right)t\right] \exp\left(2\sigma^2 t\right) \\ &= \mathbb{E}\left[X_0^2\right] \exp\left[2\left(\mu + \frac{\sigma^2}{2}\right)t\right]\end{aligned}$$

and then

$$\begin{aligned}\operatorname{var}[X_t] &= \mathbb{E}\left[X_t^2\right] - \left(\mathbb{E}[X_t]\right)^2 \\ &= \mathbb{E}\left[X_0^2\right] \exp\left[2\left(\mu + \frac{\sigma^2}{2}\right)t\right] - \left(\mathbb{E}[X_0]\right)^2 \exp(2\mu t)\end{aligned}$$

iii) See Exercise 1.13.

iv) By the definition of Y_t we have

$$Y_t = Y_0 + \left(\mu - \frac{\sigma^2}{2}\right)t + \sigma W_t$$

which is a Gaussian process with

$$\mu(t) = \mathbb{E}[Y_t] = \mathbb{E}[Y_0] + \left(\mu - \frac{\sigma^2}{2}\right)t$$

and

$$c(s,t) = \operatorname{cov}(Y_s, Y_t) = \operatorname{var}[Y_0] + \operatorname{cov}(\sigma W_s, \sigma W_t) = \operatorname{var}[Y_0] + \sigma^2 (s \wedge t)$$

□

If the initial value X_0 of a GBM is a positive constant, then by the previous proposition, the GBM takes strictly positive values, satisfying

$$Y_t = \ln X_t \sim N\left(\ln X_0 + \left(\mu - \frac{\sigma^2}{2}\right)t, \sigma^2 t\right)$$

i.e., X_t has a lognormal distribution for every $t \geq 0$.

The positivity of the GBM (with $X_0 > 0$), makes it a good candidate for modeling financial asset prices. Furthermore, its lognormal distribution allows for closed-form solutions. For these reasons, this process has been used as a basis for developing various models in quantitative finance, such as the Black-Scholes model. Figure 3.2 shows a sample path of GBM with $X_0 = 100$, $\mu = 0.05$ and $\sigma = 0.1$.

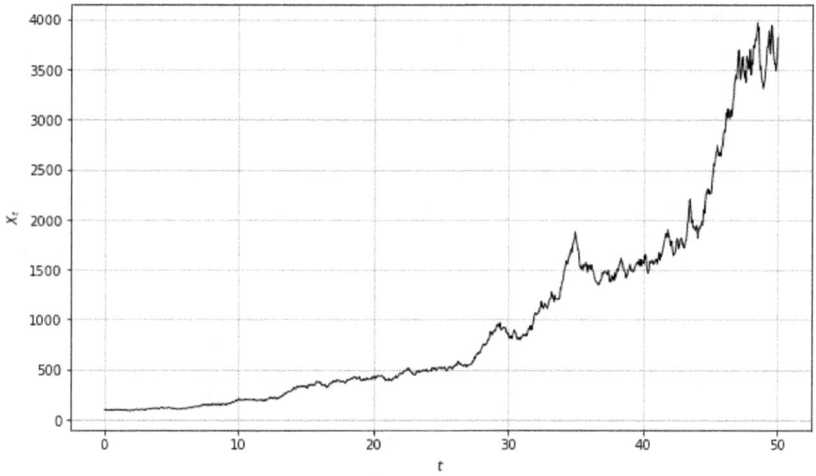

Figure 3.2: A geometric Brownian motion path.

Brownian Bridge

Let a, b be two constants. The Brownian bridge from a to b is the solution of the SDE

$$dX_t = \frac{b - X_t}{1 - t} dt + dW_t, \quad 0 \le t < 1 \tag{3.22}$$

with $X_0 = a$.

Proposition 3.24. *The unique solution of equation (3.22) is*

$$X_t = (1 - t)\, a + bt + (1 - t) \int_0^t \frac{dW_s}{1 - s}, \quad 0 \le t < 1 \tag{3.23}$$

and satisfies:

i) $\mathbb{E}\,[X_t] = (1 - t)\, a + bt$

ii) $\text{var}\,[X_t] = t\,(1 - t)$

iii) $\lim_{t \to 1^-} \mathbb{E}\,[X_t] = b$

iv) $\lim_{t \to 1^-} \text{var}\,[X_t] = 0$

v) $\lim_{t \to 1^-} X_t = b$ *a.s.*

vi) X_t *is a Gaussian process with mean and covariance functions given by*

$$\mu\,(t) = (1 - t)\, a + bt$$
$$c\,(s, t) = s \wedge t - st$$

Proof. The SDE (3.22) is a linear SDE in the narrow sense of the type (3.15), with $n = d = 1$, $\mathbf{F}(t) = -\frac{1}{1-t}$, $\mathbf{f}(t) = \frac{b}{1-t}$ and $\mathbf{g}_1(t) = 1$. The fundamental solution is obtained solving the ODE

$$\frac{d\Phi(t)}{dt} = -\frac{1}{1-t}\Phi(t), \quad \Phi(0) = 1$$

whose solution is $\Phi(t) = 1 - t$. Then, by (3.16), we find

$$\begin{aligned}
X_t &= (1-t)\left[a + \int_0^t \frac{1}{1-s}\frac{b}{1-s}ds + \int_0^t \frac{1}{1-s}dW_s\right] \\
&= (1-t)\left[a + \int_0^t \frac{b}{(1-s)^2}ds + \int_0^t \frac{dW_s}{1-s}\right] \\
&= (1-t)\left[a + b\frac{t}{1-t} + \int_0^t \frac{dW_s}{1-s}\right] \\
&= (1-t)\,a + bt + (1-t)\int_0^t \frac{dW_s}{1-s}
\end{aligned}$$

i) Immediate from (3.23).

ii) Taking variance in expression (3.23) we have

$$\begin{aligned}
\operatorname{var}[X_t] &= \operatorname{var}\left[(1-t)\int_0^t \frac{dW_s}{1-s}\right] = (1-t)^2\,\mathbb{E}\left[\left(\int_0^t \frac{dW_s}{1-s}\right)^2\right] \\
&= (1-t)^2\int_0^t \frac{ds}{(1-s)^2} = (1-t)^2\frac{t}{1-t} = t(1-t)
\end{aligned}$$

iii) and iv) are immediate from i) and ii).

v) It suffices to prove that

$$\lim_{t\to 1^-}(1-t)\int_0^t \frac{dW_s}{1-s} = 0 \text{ a.s.}$$

The proof of this result can be found in Karatzas and Shreve (1991), page 359.

vi) The Gaussian property and the expression for $\mu(t)$ are immediate from (3.23). For the covariance function we have

$$\begin{aligned}
c(s,t) = \operatorname{cov}(X_s, X_t) &= (1-s)(1-t)\operatorname{cov}\left(\int_0^s \frac{dW_u}{1-u}, \int_0^t \frac{dW_u}{1-u}\right) \\
&= (1-s)(1-t)\int_0^{s\wedge t}\frac{du}{(1-u)^2} = (1-s)(1-t)\left[\frac{1}{1-(s\wedge t)} - 1\right] \\
&= (1-s)(1-t)\left(\frac{s\wedge t}{1-(s\wedge t)}\right) = s\wedge t - st
\end{aligned}$$

□

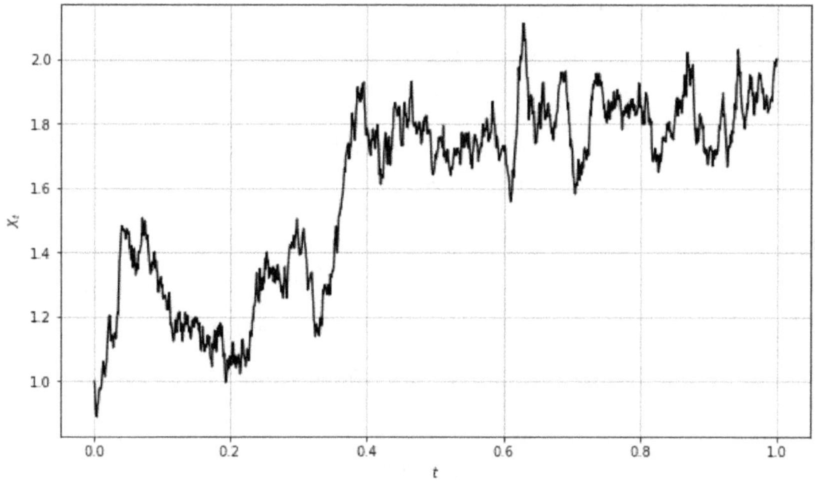

Figure 3.3: A Brownian bridge path.

The property v) allows to extend the Brownian bridge solution (3.23) from $[0, 1)$ to $[0, 1]$ by continuity, just by taking $X_1 = b$ a.s. The Brownian bridge from a to b can be seen as a Brownian motion on $[0, 1]$, started at a and conditioned to arrive at b at time 1 (Karatzas and Shreve (1991), problem 6.13, Chapter 5). From ii) and iv) it is immediate that the Brownian bridge from a to b has zero variance at the endpoints of the interval, and maximum variance $1/4$ at the midpoint. As a consequence of vi), we have that $X_t \sim N((1-t)a + bt, t(1-t))$. Figure 3.3 shows a sample path of a Brownian bridge from 1 to 2.

Exercise 3.25. Prove that the process

$$Y_t = a(1-t) + bt + (W_t - tW_1), \quad 0 \leq t < 1$$

has the same distribution that the Brownian bridge X_t of expression (3.23).

3.4 Integration Techniques

Just as in ordinary calculus, stochastic calculus provides several integration techniques that can be applied to SDEs beyond the linear case.

Transformation to Linear

The following result states sufficient conditions for a one-dimensional SDE with time-independent coefficients to be transformed into a linear SDE.

Lemma 3.26. *The SDE*

$$dX_t = \mu(X_t)\,dt + \sigma(X_t)\,dW_t$$

can be transformed into the linear SDE

$$dZ_t = (\alpha + \beta Z_t)\,dt + (\gamma + \delta Z_t)\,dW_t$$

if, and only if

$$\frac{d}{dx}\left[\frac{\frac{d}{dx}\left(\varkappa'(x)\,\sigma(x)\right)}{\varkappa'(x)}\right] = 0 \tag{3.24}$$

where

$$\varkappa(x) := \frac{\mu(x)}{\sigma(x)} - \frac{1}{2}\sigma'(x)$$

If (3.24) holds, then the transformation $Z_t = f(X_t)$ is given by

$$f(x) = \begin{cases} e^{\delta\lambda(x)} & \text{if} \quad \delta \neq 0 \\ \gamma\lambda(x) & \text{if} \quad \delta = 0 \end{cases}$$

where

$$\lambda(x) = \int_0^x \frac{dy}{\sigma(y)}$$

$$\delta = -\frac{\frac{d}{dx}\left(\varkappa'(x)\,\sigma(x)\right)}{\varkappa'(x)}$$

and $\gamma \neq 0$ is an arbitrary constant.

Proof. See Schilling and Partzsch (2012), Lemma 18.8. □

The previous lemma cannot be applied when $\varkappa'(x) = 0$. For this case we have the following result.

Lemma 3.27. *If $\varkappa'(x) = 0$, then the SDE*

$$dX_t = \mu(X_t)\,dt + \sigma(X_t)\,dW_t$$

can be transformed into the SDE

$$dZ_t = c\,dt + dW_t$$

where $Z_t = \lambda(X_t) = \int_0^{X_t} \frac{dy}{\sigma(y)}$, and c is the constant given by

$$c = \varkappa(x) = \frac{\mu(x)}{\sigma(x)} - \frac{1}{2}\sigma'(x).$$

Proof. By Itô's formula

$$dZ_t = \left[\frac{\mu(X_t)}{\sigma(X_t)} - \frac{1}{2}\sigma'(X_t)\right]dt + dW_t = c\,dt + dW_t$$

which completes the proof. □

Exercise 3.28. Find the solution of the SDE

$$dX_t = \left(\sqrt{1 + X_t^2} + \frac{1}{2}X_t\right)dt + \sqrt{1 + X_t^2}\,dW_t, \quad X_0 = 0 \tag{3.25}$$

Exact SDEs

As in the case of ODEs, if it is possible to find a process such that the SDE can be expressed as the Itô differential of that process, then the SDE is said to be exact and its solution can be obtained immediately. In this section, we restrict ourselves to SDEs whose coefficients can be expressed as functions of t and W_t.

Definition 3.29. (Exact SDE) The SDE

$$dX_t = \mu(t, W_t)\, dt + \sigma(t, W_t)\, dW_t \tag{3.26}$$

is called exact if there exists a function $f : [0, \infty) \times \mathbb{R} \to \mathbb{R}$ of class $C^{1,2}$ such that

$$\mu(t,x) = \frac{\partial f}{\partial t}(t,x) + \frac{1}{2}\frac{\partial^2 f}{\partial x^2}(t,x) \tag{3.27}$$

and

$$\sigma(t,x) = \frac{\partial f}{\partial x}(t,x) \tag{3.28}$$

Proposition 3.30. *If the SDE (3.26) is exact, then every solution of (3.26) is of the form*

$$X_t = f(t, W_t) + C$$

where f is the function that satisfies (3.27)-(3.28), and C is a constant.

Proof. Using (3.27) and (3.28), we can rewrite (3.26) as

$$dX_t = \left[\frac{\partial f}{\partial t}(t, W_t) + \frac{1}{2}\frac{\partial^2 f}{\partial x^2}(t, W_t)\right] dt + \frac{\partial f}{\partial x}(t, W_t)\, dW_t$$

and using Itô's formula for Brownian motion, equation (2.25), we obtain

$$dX_t = d\left[f(t, W_t)\right]$$

which proves the proposition. □

In the applications, before trying to find a solution of the system (3.27), (3.28), it is convenient to use the following necessary condition for an SDE to be exact.

Proposition 3.31. *If the SDE (3.26) is exact, then the functions μ and σ must satisfy the condition*

$$\frac{\partial \mu}{\partial x}(t,x) = \frac{\partial \sigma}{\partial t}(t,x) + \frac{1}{2}\frac{\partial^2 \sigma}{\partial x^2}(t,x) \tag{3.29}$$

Proof. Differentiating (3.27) with respect to x we have

$$\begin{aligned}
\frac{\partial \mu}{\partial x}(t,x) &= \frac{\partial}{\partial x}\left[\frac{\partial f}{\partial t}(t,x)\right] + \frac{1}{2}\frac{\partial^3 f}{\partial x^3}(t,x) \\
&= \frac{\partial}{\partial t}\left[\frac{\partial f}{\partial x}(t,x)\right] + \frac{1}{2}\frac{\partial^2}{\partial x^2}\left[\frac{\partial f}{\partial x}(t,x)\right] \\
&= \frac{\partial \sigma}{\partial t}(t,x) + \frac{1}{2}\frac{\partial^2 \sigma}{\partial x^2}(t,x)
\end{aligned}$$

where we have used Schwarz's Theorem and (3.28). □

Example 3.32. The SDE

$$dX_t = \left(1 + W_t^2\right) dt + \left(t^4 + W_t^2\right) dW_t$$

is not exact. To see this, it is enough to consider that in this case we have $\mu(t,x) = 1 + x^2$, $\sigma(t,x) = t^4 + x^2$. Therefore

$$\frac{\partial \mu}{\partial x}(t,x) = 2x$$
$$\frac{\partial \sigma}{\partial t}(t,x) + \frac{1}{2}\frac{\partial^2 \sigma}{\partial x^2}(t,x) = 4t^3 + 1$$

and then, the equality (3.29) is not satisfied. □

Exercise 3.33. Solve the SDE

$$dX_t = 2tW_t dt + \left(t^2 + W_t\right) dW_t, \quad X_0 = 0 \tag{3.30}$$

Integrating Factors

Sometimes it is possible to solve an SDE by finding a factor, known as *integrating factor*, that converts it into an exact SDE, which can be solved more easily.

Proposition 3.34. *Consider the SDE*

$$dX_t = f(t, X_t)\, dt + g(t)\, X_t dW_t, \quad X_0 = x$$

where $f : [0,\infty) \times \mathbb{R} \to \mathbb{R}$ *and* $g : [0,\infty) \to \mathbb{R}$ *are deterministic functions, and the process*

$$\rho_t := \exp\left\{\frac{1}{2}\int_0^t g^2(s)\, ds - \int_0^t g(s)\, dW_s\right\}$$

Then, if we define $Y_t := \rho_t X_t$, Y_t *satisfies the ODE*

$$dY_t = \rho_t f\left(t, \frac{Y_t}{\rho_t}\right) dt \tag{3.31}$$

Proof. Applying Itô's formula to ρ_t we obtain

$$d\rho_t = \frac{1}{2}\rho_t g^2(t)\, dt + \rho_t\left[-g(t)\, dW_t + \frac{1}{2}g^2(t)\, dt\right] = \rho_t\left[g^2(t)\, dt - g(t)\, dW_t\right]$$

and then, by integration by parts

$$\begin{aligned} dY_t &= d(\rho_t X_t) = \rho_t dX_t + X_t d\rho_t + d[\rho_\cdot, X_\cdot]_t \\ &= \rho_t [f(t, X_t)\, dt + g(t)\, X_t dW_t] + X_t\rho_t\left[g^2(t)\, dt - g(t)\, dW_t\right] \\ &- \rho_t X_t g^2(t)\, dt \\ &= \rho_t f(t, X_t)\, dt = \rho_t f\left(t, \frac{Y_t}{\rho_t}\right) dt \end{aligned}$$

□

Exercise 3.35. Solve the SDE

$$dX_t = \frac{1}{X_t}dt + \alpha X_t dW_t$$

where α is a constant and $X_0 > 0$, by the method of integrating factors.

3.5 Markov and Strong Markov Properties of the Solutions

Stochastic differential equations provide a method for creating new stochastic processes which satisfy the Markov property, as stated in the following result.

Theorem 3.36. *Consider the SDE*

$$d\mathbf{X}_t = \mathbf{b}(t, \mathbf{X}_t)\, dt + \boldsymbol{\Sigma}(t, \mathbf{X}_t)\, d\mathbf{W}_t, \quad 0 \le t \le T, \;\; \mathbf{X}_0 = \xi \tag{3.32}$$

and assume that it satisfies the hypotheses of Theorem (3.4). Then, its unique solution is a Markov process.

Proof. For fixed $s \in [0, T]$ and $\mathbf{x} \in \mathbb{R}^n$, let $\mathbf{X}_t^{s,\mathbf{x}}$ denote the unique solution of the SDE

$$d\mathbf{X}_t = \mathbf{b}(t, \mathbf{X}_t)\, dt + \boldsymbol{\Sigma}(t, \mathbf{X}_t)\, d\mathbf{W}_t, \quad s \le t \le T, \;\; \mathbf{X}_s = \mathbf{x}$$

Since $\mathbf{x}$ is deterministic, $\mathbf{X}_t^{s,\mathbf{x}}$ depends only on the Brownian increments $(\mathbf{W}_u - \mathbf{W}_s)_{u \ge s}$ (see Kuo (2006), Section 10.6, for a more detailed argument). Therefore, $\mathbf{X}_t^{s,\mathbf{x}}$ is independent of $\mathcal{F}_s$. Thus, for any $\mathcal{F}_s$-measurable random variable $\mathbf{Z}$, we have

$$\mathbb{P}\left[\mathbf{X}_t^{s,\mathbf{Z}} \in B \,|\mathcal{F}_s\right] = \mathbb{P}[\mathbf{X}_t^{s,\mathbf{x}} \in B]\big|_{\mathbf{x}=\mathbf{Z}}$$

for any Borel set $B \in \mathcal{B}(\mathbb{R}^n)$, where the right-hand side denotes the function $\mathbf{x} \mapsto \mathbb{P}\left[\mathbf{X}_t^{s,\mathbf{x}} \in B\right]$ evaluated at $\mathbf{x} = \mathbf{Z}$. In particular, taking $\mathbf{Z} = \mathbf{X}_s$, we get

$$\mathbb{P}\left[\mathbf{X}_t^{s,\mathbf{X}_s} \in B \,|\mathcal{F}_s\right] = \mathbb{P}[\mathbf{X}_t^{s,\mathbf{x}} \in B]\big|_{\mathbf{x}=\mathbf{X}_s} \quad \forall B \in \mathcal{B}(\mathbb{R}^n) \tag{3.33}$$

By uniqueness of the solution, we have that $\mathbf{X}_t = \mathbf{X}_t^{s,\mathbf{X}_s}$ a.s. for all $t \ge s$, and then, we can rewrite (3.33) as

$$\mathbb{P}[\mathbf{X}_t \in B \,|\mathcal{F}_s] = \mathbb{P}[\mathbf{X}_t^{s,\mathbf{x}} \in B]\big|_{\mathbf{x}=\mathbf{X}_s} \forall t \ge s, \;\; \forall B \in \mathcal{B}(\mathbb{R}^n)$$

which is measurable with respect to the σ-algebra generated by $\mathbf{X}_s$, $\sigma(\mathbf{X}_s)$. Therefore, taking into account that $\sigma(\mathbf{X}_s) \subset \mathcal{F}_s$, we can write

$$\begin{aligned}
\mathbb{P}[\mathbf{X}_t \in B \,|\mathcal{F}_s] &= \mathbb{E}\{\mathbb{P}[\mathbf{X}_t \in B \,|\mathcal{F}_s] \,|\mathbf{X}_s\} \\
&= \mathbb{E}\left\{\mathbb{E}\left[\mathbf{1}_{\{\mathbf{X}_t \in B\}} \,|\mathcal{F}_s\right] |\mathbf{X}_s\right\} \\
&= \mathbb{E}\left\{\mathbf{1}_{\{\mathbf{X}_t \in B\}} \,|\mathbf{X}_s\right\} \\
&= \mathbb{P}[\mathbf{X}_t \in B \,|\mathbf{X}_s]
\end{aligned}$$

which is the condition for the multidimensional process $\mathbf{X}_t$ to be a Markov process (see, for example, Kallianpur and Sundar (2014)). □

The solutions of SDEs exhibit, like Brownian motion, the more general property of being strong Markov processes.

Theorem 3.37. *Under the assumptions of Theorem (3.36),* $\mathbf{X}_t$ *is a strong Markov process.*

Proof. See Oksendal (1998), Theorem 7.2.4 or Friedman (2006), Chapter 5, Theorem 3.4. □

The Markov and strong Markov properties of solutions of SDEs allow us to deepen our understanding of them by applying techniques from the theory of Markov processes. We begin by defining some concepts.

Definition 3.38. (Transition probability function) Given a Markov process $\mathbf{X}_t$, its transition probability function, $P(s,\mathbf{x}\,|t,B)$, is defined by

$$P(s,\mathbf{x}\,|t,B) := \mathbb{P}\left[\mathbf{X}_t \in B\,|\mathbf{X}_s = \mathbf{x}\right], \quad s<t, \quad B\in\mathcal{B}\left(\mathbb{R}^n\right)$$

and

$$P(t,\mathbf{x}\,|t,B) = \mathbf{1}_B(\mathbf{x})$$

i.e., it is the conditional distribution function of the random variable $\mathbf{X}_t$, given that the process $\mathbf{X}$ takes the value $\mathbf{x}$ at time s. In other words, $P(s,\mathbf{x}\,|t,B)$ represents the probability that the stochastic process $\mathbf{X}$ takes values in B at time t, knowing that its value at time s is $\mathbf{x}$, with $s<t$.

Remark 3.39. When working with one-dimensional Markov processes, we will use the following slightly different notation

$$P(s,x\,|t,y) := P(s,x\,|t,(-\infty,y]) = \mathbb{P}\left[X_t \le y\,|X_s = x\right]$$

Transition probability functions allows us to obtain unconditional probabilities related to solutions of SDEs, as the following result states.

Corollary 3.40. *Let* $\mathbf{X}_t$ *be the solution of the SDE*

$$d\mathbf{X}_t = \mathbf{b}(t,\mathbf{X}_t)\,dt + \mathbf{\Sigma}(t,\mathbf{X}_t)\,d\mathbf{W}_t, \quad 0\le t\le T, \ \mathbf{X}_0=\xi$$

Then, its transition probability function, $P(s,\mathbf{x}\,|t,B)$, *satisfies*

$$P(s,\mathbf{x}\,|t,B) = \mathbb{P}\left[\mathbf{X}_t^{s,\mathbf{x}} \in B\right]$$

where $\mathbf{X}_t^{s,\mathbf{x}}$ *is the solution of the SDE*

$$d\mathbf{X}_t = \mathbf{b}(t,\mathbf{X}_t)\,dt + \mathbf{\Sigma}(t,\mathbf{X}_t)\,d\mathbf{W}_t, \quad s\le t\le T, \ \mathbf{X}_s=\mathbf{x}$$

Proof. Immediate from the definition of the transition probability function, the definiton of $\mathbf{X}_t^{s,\mathbf{x}}$, and Theorem 3.36. □

Theorem 3.41. *The transition probability function of a Markov process* $\mathbf{X}_t$ *satisfies the Chapman-Kolmogorov equation*

$$P(s,\mathbf{x}\,|t,B) = \int_{\mathbb{R}^n} P(u,\mathbf{y}\,|t,B)\,P(s,\mathbf{x}\,|u,d\mathbf{y})$$

for all $u\in(s,t)$, $\mathbf{x}\in\mathbb{R}^n$ *and* $B\in\mathcal{B}\left(\mathbb{R}^n\right)$.

Proof. Applying Definition 3.38, the law of total probability, and the Markov property of $\mathbf{X}$, we have

$$\begin{aligned} P(s,\mathbf{x}\,|t,B) &= \mathbb{P}[\mathbf{X}_t \in B\,|\mathbf{X}_s = \mathbf{x}] \\ &= \int_{\mathbb{R}^n} \mathbb{P}\left[\mathbf{X}_t \in B\,|\mathbf{X}_u = \mathbf{y}, \mathbf{X}_s = \mathbf{x}\right] \mathbb{P}[\mathbf{X}_u \in d\mathbf{y}\,|\mathbf{X}_s = \mathbf{x}] \\ &= \int_{\mathbb{R}^n} \mathbb{P}[\mathbf{X}_t \in B\,|\mathbf{X}_u = \mathbf{y}]\, \mathbb{P}[\mathbf{X}_u \in d\mathbf{y}\,|\mathbf{X}_s = \mathbf{x}] \\ &= \int_{\mathbb{R}^n} P(u,\mathbf{y}\,|t,B)\, P(s,\mathbf{x}\,|u,d\mathbf{y}) \end{aligned}$$

□

In intuitive terms, the Chapman-Kolmogorov equation tells us that the transition probability function of a Markov process $\mathbf{X}$ can be obtained as the product of the transition probability from the initial instant s with $\mathbf{X}_s = \mathbf{x}$, to any arbitrary intermediate time u, where it takes the value $\mathbf{y}$, by the transition probability from this intermediate time to the final instant t, taking values in B. Finally, the product is integrated for all possible intermediate values $\mathbf{y}$ of the process at time u.

Remark 3.42. By Theorem 3.41, if $\mathbf{X}$ is a Markov process, then its transition probabilities satisfy the Chapman-Kolmogorov equation. The converse is also true: given a set of probabilities $\{P(s,\mathbf{x}\,|t,\cdot) : 0 \le s \le t \le T, \mathbf{x} \in \mathbb{R}\}$ satisfying the Chapman-Kolmogorov equation, then there exists a Markov process such that $P(s,\mathbf{x}\,|t,B)$ is its transition probability function (Kuo (2006), Theorem 10.5.14).

Exercise 3.43. Find the transition probability function of one-dimensional Brownian motion, and verify that it satisfies the Chapman-Kolmogorov equation.

Definition 3.44. (Transition operator) Let $\mathbf{X}_t$ be the solution of the SDE (3.32) and $P(s,\mathbf{x}\,|t,\mathbf{y})$ its associated transition probability function. For any s,t with $0 \le s \le t \le T$ and any measurable function $f : \mathbb{R}^n \to \mathbb{R}$, we define the transition operator T_s^t by

$$\left(T_s^t f\right)(\mathbf{x}) := \int_{\mathbb{R}^n} f(\mathbf{y})\, P(s,\mathbf{x}\,|t,d\mathbf{y}) = \mathbb{E}\left[f\left(\mathbf{X}_t^{s,\mathbf{x}}\right)\right] \tag{3.34}$$

and

$$\left(T_s^s f\right)(\mathbf{x}) := f(\mathbf{x}) \tag{3.35}$$

assuming the integral in (3.34) exists.

Proposition 3.45. *For any $u \in [s,t]$, the following identity holds*

$$T_s^u T_u^t = T_s^t$$

Proof. For any $\mathbf{x} \in \mathbb{R}^n$ and any measurable function $f : \mathbb{R}^n \to \mathbb{R}$, applying (3.34) and the Chapman-Kolmogorov equation, we have the following chain of equalities that prove the result

$$
\begin{aligned}
\left(T_s^u T_u^t f\right)(\mathbf{x}) &= T_s^u \left[\int_{\mathbb{R}^n} f(\mathbf{y})\, P(u, \mathbf{x} \,|t, d\mathbf{y})\right] \\
&= \int_{\mathbb{R}^n} \left[\int_{\mathbb{R}^n} f(\mathbf{y})\, P(u, \mathbf{v} \,|t, d\mathbf{y})\right] P(s, \mathbf{x} \,|u, d\mathbf{v}) \\
&= \int_{\mathbb{R}^n} f(\mathbf{y})\, P(s, \mathbf{x} \,|t, d\mathbf{y}) \\
&= \left(T_s^t f\right)(\mathbf{x})
\end{aligned}
$$

□

This property of the transition operators motivates the following definition.

Definition 3.46. (Semigroup of a Markov process) The family of operators $\{T_s^t\}$, $0 \le s \le t \le T$, defined by (3.34), (3.35) is called the semigroup of the Markov process $\mathbf{X}_t$.[3]

Let $C_b^2(\mathbb{R}^n)$ be the set of twice continuously differentiable functions $f : \mathbb{R}^n \to \mathbb{R}$ which are bounded and have bounded partial derivatives up to order 2.

Theorem 3.47. *Let $\mathbf{X}_t$ be the solution of the SDE (3.32) with bounded coefficients, and consider the operator $\mathcal{A}_t$, defined on the set $C_b^2(\mathbb{R}^n)$ by*

$$
(\mathcal{A}_t f)(\mathbf{x}) = \frac{1}{2}\sum_{i,j=1}^{n} a_{ij}(t, \mathbf{x}) \frac{\partial^2 f}{\partial x_i \partial x_j}(\mathbf{x}) + \sum_{i=1}^{n} b_i(t, \mathbf{x}) \frac{\partial f}{\partial x_i}(\mathbf{x}) \tag{3.36}
$$

where $a_{ij} = \left(\mathbf{\Sigma}\mathbf{\Sigma}^\top\right)_{ij}$ and b_i is the i-th component of $\mathbf{b}$. Then

$$
\left(T_s^t f\right)(\mathbf{x}) = f(\mathbf{x}) + \int_s^t \left[T_s^u (\mathcal{A}_u f)\right](\mathbf{x})\, du, \quad 0 \le s < t \le T \tag{3.37}
$$

for all $\mathbf{x} \in \mathbb{R}^n$.

Proof. By the definition of $\mathbf{X}_t^{s,\mathbf{x}}$, we have

$$
\mathbf{X}_t^{s,\mathbf{x}} = \mathbf{x} + \int_s^t \mathbf{b}\left(u, \mathbf{X}_u^{s,\mathbf{x}}\right) du + \int_s^t \mathbf{\Sigma}\left(u, \mathbf{X}_u^{s,\mathbf{x}}\right) d\mathbf{W}_u
$$

Applying the multidimensional Itô formula (2.38) we get

$$
f\left(\mathbf{X}_t^{s,\mathbf{x}}\right) = f(\mathbf{x}) + \int_s^t (\mathcal{A}_u f)\left(\mathbf{X}_u^{s,\mathbf{x}}\right) du + \sum_{i,j=1}^{n} \int_s^t \sigma_{ij}\left(u, \mathbf{X}_u^{s,\mathbf{x}}\right) \frac{\partial f}{\partial x_i}\left(\mathbf{X}_u^{s,\mathbf{x}}\right) dW_u^{(j)} \tag{3.38}
$$

[3] Given a set S, a family of mappings $U_r^t : S \to S$, $r \le t$, is known as a semigroup with two parameters (or propagator) if U_t^t is the identity and $U_s^t U_r^s = U_r^t$ for $r \le s \le t$ (Kolokoltsov, 2011).

where $W_u^{(j)}$ is the j-th component of $\mathbf{W}$. By the assumptions, the stochastic integrals are martingales. Then, taking expectations and applying the martingale property we obtain

$$\mathbb{E}\left[f\left(\mathbf{X}_t^{s,\mathbf{x}}\right)\right] = f(\mathbf{x}) + \int_s^t \mathbb{E}\left[\left(\mathcal{A}_u f\right)\left(\mathbf{X}_u^{s,\mathbf{x}}\right)\right] du \tag{3.39}$$

which can be rewritten as

$$\left(T_s^t f\right)(\mathbf{x}) = f(\mathbf{x}) + \int_s^t \left[T_s^u\left(\mathcal{A}_u f\right)\right](\mathbf{x})\, du$$

□

Definition 3.48. (Infinitesimal generator of a semigroup) The operator defined by

$$\left(\mathcal{T}_s f\right)(\mathbf{x}) = \lim_{t \to s} \frac{\left(T_s^t f\right)(\mathbf{x}) - f(\mathbf{x})}{t - s}$$

is known as the infinitesimal generator of the semigroup $\left\{T_s^t\right\}$, assumed that the limit of the right-hand side exists for every $\mathbf{x} \in \mathbb{R}^n$.

Corollary 3.49. *If the coefficients* $\mathbf{b}(t, \mathbf{x})$ *and* $\mathbf{\Sigma}(t, \mathbf{x})$ *are continuous in* t*, then* $\mathcal{A}_t$ *given by (3.36) is the infinitesimal generator of the semigroup of the solution of (3.32).*

Proof. From equation (3.37) we have

$$\begin{aligned}\lim_{t \to s} \frac{\left(T_s^t f\right)(\mathbf{x}) - f(\mathbf{x})}{t - s} &= \lim_{t \to s} \frac{1}{t - s} \int_s^t \left[T_s^u\left(\mathcal{A}_u f\right)\right](\mathbf{x})\, du \\ &= \left[T_s^s\left(\mathcal{A}_s f\right)\right](\mathbf{x}) = \left(\mathcal{A}_s f\right)(\mathbf{x})\end{aligned}$$

□

Exercise 3.50. Find the infinitesimal generators asociated to the mean-reverting Ornstein-Uhlenbeck process and the geometric Brownian motion.

Expression (3.39) in the proof of Theorem 3.47 has a special status as stated in the following result.

Corollary 3.51. ***(Dynkin's formula)*** *Under the hypotheses of Theorem 3.47 we have*

$$\mathbb{E}\left[f\left(\mathbf{X}_t^{s,\mathbf{x}}\right)\right] = f(\mathbf{x}) + \int_s^t \mathbb{E}\left[\left(\mathcal{A}_u f\right)\left(\mathbf{X}_u^{s,\mathbf{x}}\right)\right] du$$

Remark 3.52. Dynkin's formula also holds if t is replaced by a stopping time τ with $\mathbb{E}[\tau] < \infty$ (see Oksendal (1998), Theorem 7.4.1).

The following result is another consequence of Theorem 3.47.

Corollary 3.53. *Under the hypotheses of Theorem 3.47, the process*

$$M_t = f\left(\mathbf{X}_t^{s,\mathbf{x}}\right) - f(\mathbf{x}) - \int_s^t \left(\mathcal{A}_u f\right)\left(\mathbf{X}_u^{s,\mathbf{x}}\right) du$$

is a martingale.

Proof. The proof is immediate from expression (3.38). □

3.6 Diffusions

According to Kuo (2006), Kiyoshi Itô's main motivation for developing stochastic calculus was to obtain diffusions from solution to SDEs. To determine the specific relationship between these two concepts, we first need to define a diffusion.

Definition 3.54. (Diffusion) A Markov process $\mathbf{X}_t$ is a diffusion if it has continuous paths and there exist two functions, $\boldsymbol{\rho}(t, \mathbf{x})$ and $\mathbf{Q}(t, \mathbf{x})$, with values in $\mathbb{R}^n$ and $\mathbb{R}^{n \times n}$, respectively, such that its transition probability function, $P(s, \mathbf{x} | t, B)$, satisfies, for any $c > 0$, the following conditions:

i) $\lim_{\varepsilon \to 0^+} \frac{1}{\varepsilon} \int_{\|\mathbf{y}-\mathbf{x}\| \geq c} P(t, \mathbf{x} | t + \varepsilon, d\mathbf{y}) = 0$

ii) $\lim_{\varepsilon \to 0^+} \frac{1}{\varepsilon} \int_{\|\mathbf{y}-\mathbf{x}\| < c} (y_i - x_i) P(t, \mathbf{x} | t + \varepsilon, d\mathbf{y}) = \rho_i(t, \mathbf{x}) \quad \forall i \in \{1, \ldots, n\}$

iii) $\lim_{\varepsilon \to 0^+} \frac{1}{\varepsilon} \int_{\|\mathbf{y}-\mathbf{x}\| < c} (y_i - x_i)(y_j - x_j) P(t, \mathbf{x} | t + \varepsilon, d\mathbf{y}) = q_{ij}(t, \mathbf{x})$
$\forall i, j \in \{1, \ldots, n\}$.

The vector

$$\boldsymbol{\rho}(t, \mathbf{x}) = (\rho_1(t, \mathbf{x}), \ldots, \rho_n(t, \mathbf{x}))$$

and the matrix

$$\mathbf{Q}(t, \mathbf{x}) = \left[q_{ij}(t, \mathbf{x})\right]_{n \times n}$$

are called, respectively, the drift and the diffusion coefficients of the diffusion $\mathbf{X}_t$.

Remark 3.55. The inequality $\|\mathbf{y} - \mathbf{x}\| \geq c$ on the integral in condition i) is set to prevent X_t from having instantaneous jumps.

Remark 3.56. Let $c_1 > c$ and consider the limit

$$\begin{aligned}
\lim_{\varepsilon \to 0^+} \frac{1}{\varepsilon} \int_{\|\mathbf{y}-\mathbf{x}\| < c_1} & (y_i - x_i) P(t, \mathbf{x} | t + \varepsilon, d\mathbf{y}) \\
&= \lim_{\varepsilon \to 0^+} \frac{1}{\varepsilon} \int_{\|\mathbf{y}-\mathbf{x}\| < c} (y_i - x_i) P(t, \mathbf{x} | t + \varepsilon, d\mathbf{y}) \\
&+ \lim_{\varepsilon \to 0^+} \frac{1}{\varepsilon} \int_{c \leq \|\mathbf{y}-\mathbf{x}\| < c_1} (y_i - x_i) P(t, \mathbf{x} | t + \varepsilon, d\mathbf{y})
\end{aligned}$$

For the second limit in the right-hand side, we have

$$\begin{aligned}
\lim_{\varepsilon \to 0^+} \frac{1}{\varepsilon} \int_{c \leq \|\mathbf{y}-\mathbf{x}\| < c_1} & (y_i - x_i) P(t, \mathbf{x} | t + \varepsilon, d\mathbf{y}) \\
&\leq \lim_{\varepsilon \to 0^+} \frac{1}{\varepsilon} \int_{c \leq \|\mathbf{y}-\mathbf{x}\| < c_1} |y_i - x_i| P(t, \mathbf{x} | t + \varepsilon, d\mathbf{y}) \\
&\leq c_1 \lim_{\varepsilon \to 0^+} \frac{1}{\varepsilon} \int_{\|\mathbf{y}-\mathbf{x}\| \geq c} P(t, \mathbf{x} | t + \varepsilon, d\mathbf{y}) = 0
\end{aligned}$$

where in the last step we have used i) of Definition 3.54. We have proved that for any $c_1 > c > 0$

$$\begin{aligned}\lim_{\varepsilon\to 0^+} \frac{1}{\varepsilon} \int_{\|\mathbf{y}-\mathbf{x}\|<c_1} (y_i - x_i)\, P(t, \mathbf{x} \,| t+\varepsilon, d\mathbf{y}) \\ = \lim_{\varepsilon\to 0^+} \frac{1}{\varepsilon} \int_{\|\mathbf{y}-\mathbf{x}\|<c} (y_i - x_i)\, P(t, \mathbf{x} \,| t+\varepsilon, d\mathbf{y}) \\ = \rho_i(t, \mathbf{x}) \end{aligned} \tag{3.40}$$

and then, the limit in ii) of the same definition is independent of the value of the constant c. The same occurs with the limit in condition iii).

Regarding the conditions ii) and iii) of Definition 3.54, we have the following result.

Lemma 3.57. *Conditions ii) and iii) of Definition 3.54 can be rewritten as*

ii)' $\rho_i(t, \mathbf{x}) = \lim_{\varepsilon\to 0^+} \frac{1}{\varepsilon} \mathbb{E}\left[X_{t+\varepsilon}^{(i)} - X_t^{(i)} \,|\mathbf{X}_t = \mathbf{x}\right]$

iii)' $q_{ij}(t, \mathbf{x}) = \lim_{\varepsilon\to 0^+} \frac{1}{\varepsilon} \mathbb{E}\left[\left(X_{t+\varepsilon}^{(i)} - X_t^{(i)}\right)\left(X_{t+\varepsilon}^{(j)} - X_t^{(j)}\right) |\mathbf{X}_t = \mathbf{x}\right]$

where $X_t^{(i)}$ *is the i-th component of the process* $\mathbf{X}_t$.

Proof. By Remark 3.56 we can rewrite (3.40) in the form

$$\begin{aligned}\rho_i(t, \mathbf{x}) &= \lim_{\varepsilon\to 0^+} \frac{1}{\varepsilon} \int_{\mathbb{R}^n} (y_i - x_i)\, P(t, \mathbf{x} \,| t+\varepsilon, d\mathbf{y}) \\ &= \lim_{\varepsilon\to 0^+} \frac{1}{\varepsilon} \mathbb{E}\left[X_{t+\varepsilon}^{(i)} - X_t^{(i)} \,|\mathbf{X}_t = \mathbf{x}\right]\end{aligned}$$

and similarly, for the diffusion coefficient

$$\begin{aligned}q_{ij}(t, \mathbf{x}) &= \lim_{\varepsilon\to 0^+} \frac{1}{\varepsilon} \int_{\mathbb{R}^n} (y_i - x_i)\left(y_j - x_j\right) P(t, \mathbf{x} \,| t+\varepsilon, d\mathbf{y}) \\ &= \lim_{\varepsilon\to 0^+} \frac{1}{\varepsilon} \mathbb{E}\left[\left(X_{t+\varepsilon}^{(i)} - X_t^{(i)}\right)\left(X_{t+\varepsilon}^{(j)} - X_t^{(j)}\right) |\mathbf{X}_t = \mathbf{x}\right]\end{aligned}$$

□

Now, Lemma 3.57 allows us to interpret ii) and iii) of Definition 3.54. By ii)', the i-th component of the drift coefficient of $\mathbf{X}_t$, $\rho_i(t, \mathbf{x})$, is the instantaneous rate of change of the mean of $X_t^{(i)}$ conditional on $\mathbf{X}_t = \mathbf{x}$. Analogously, by iii)', the diffusion coefficients, $q_{ij}(t, \mathbf{x})$, are the instantaneous rates of change of the covariance of $X_t^{(i)}$ and $X_t^{(j)}$, conditional on $\mathbf{X}_t = \mathbf{x}$.

Exercise 3.58. Prove that one-dimensional Brownian motion, W_t, is a diffusion and find its drift and diffusion coefficients.

The following result states that, under some conditions on the coefficients, the solutions of SDEs are diffusions, and provides a direct way of obtaining the drift and diffusion coefficients without calculating the limits in ii)' and iii)' of Lemma 3.57.

Theorem 3.59. *If the SDE*

$$d\mathbf{X}_t = \mathbf{b}(t, \mathbf{X}_t)\, dt + \boldsymbol{\Sigma}(t, \mathbf{X}_t)\, d\mathbf{W}_t, \quad 0 \le t \le T, \quad \mathbf{X}_0 = \xi$$

satisfies the hypotheses of Theorem 3.4, and $\mathbf{b}$ *and* $\boldsymbol{\Sigma}$ *are continuous on* $[0,T] \times \mathbb{R}^n$, *then, the solution is a diffusion with drift coefficient* $\boldsymbol{\rho}(t,\mathbf{x}) = \mathbf{b}(t,\mathbf{x})$ *and diffusion coefficient* $\mathbf{Q}(t,\mathbf{x}) = \boldsymbol{\Sigma}(t,\mathbf{x})\boldsymbol{\Sigma}^{\top}(t,\mathbf{x})$.

Proof. See Kuo (2006), page 215. □

Example 3.60. By Theorem 3.59, it is immediate to obtain that the mean-reverting Ornstein-Uhlenbeck process of equation (3.17) is a diffusion, with drift coefficient $\rho(t,x) = -\alpha(x-\mu)$ and diffusion coefficient $q(t,x) = \sigma^2$.

3.7 Kolmogorov's Equations and Feynman-Kac Formula

In Section 3.5, we had a first look at the relationship between Markov processes and PDEs, through Theorem 3.47, which relates transition operators to the partial differential operator $\mathcal{A}_u$. In this section, we delve deeper into this relationship studying certain PDEs (the Kolmogorov's PDEs) that are satisfied by the transition probabilities of Markov processes, and a probabilistic representation of the solutions of some specific PDEs (the Feynman-Kac formula). We will focus on the case of diffusions generated by SDEs, since it is the case that appears most frequently in applications.

Kolmogorov's PDEs

In this subsection we will work with a diffusion $\mathbf{X}_t$, solution of the SDE (3.1), satisfying the hypotheses of Theorem 3.59. We assume that the transition probabilities admit a density and thus we can write

$$P(s,\mathbf{x}\,|t,d\mathbf{y}) = p(s,\mathbf{x}\,|t,\mathbf{y})\, d\mathbf{y}$$

We also assume that the functions $(s,\mathbf{x}) \mapsto p(s,\mathbf{x}\,|t,\mathbf{y})$ for any fixed $(t,\mathbf{y})$ and $(t,\mathbf{y}) \mapsto p(s,\mathbf{x}\,|t,\mathbf{y})$ for any fixed $(s,\mathbf{x})$, are of class $C^{1,2}$. We will refer to variables $(s,\mathbf{x})$ as *backward variables*, and $(t,\mathbf{y})$ as *forward variables*.

Theorem 3.61. ***(Kolmogorov's PDEs)*** *The transition probability density* $p(s,\mathbf{x}\,|t,\mathbf{y})$ *satisfies the PDE in the backward variables*

$$\frac{\partial}{\partial s} p(s,\mathbf{x}\,|t,\mathbf{y}) + \sum_{i=1}^{n} \rho_i(s,\mathbf{x}) \frac{\partial}{\partial x_i} p(s,\mathbf{x}\,|t,\mathbf{y}) + \frac{1}{2}\sum_{i,j=1}^{n} q_{ij}(s,\mathbf{x}) \frac{\partial^2}{\partial x_i \partial x_j} p(s,\mathbf{x}\,|t,\mathbf{y}) = 0, \quad s < t \tag{3.41}$$

$$p(t,\mathbf{x}\,|t,\mathbf{y}) = \delta(\mathbf{y}-\mathbf{x}) \tag{3.42}$$

which is called the Kolmogorov backward PDE, and the PDE in the forward variables

$$\frac{\partial}{\partial t} p(s,\mathbf{x}\,|t,\mathbf{y}) + \sum_{i=1}^{n} \frac{\partial}{\partial y_i} \left[p(s,\mathbf{x}\,|t,\mathbf{y})\, \rho(t,\mathbf{y}) \right]$$

$$- \frac{1}{2} \sum_{i,j=1}^{n} \frac{\partial^2}{\partial y_i \partial y_j} \left[p(s,\mathbf{x}\,|t,\mathbf{y})\, q_{ij}(t,\mathbf{y}) \right] = 0, \quad t > s \tag{3.43}$$

$$p(s,\mathbf{x}\,|s,\mathbf{y}) = \delta(\mathbf{y} - \mathbf{x}) \tag{3.44}$$

which is called the Kolmogorov forward PDE or Fokker-Planck equation.[4]

Proof. For $s < t$ and $\varepsilon > 0$, with $s < s + \varepsilon < t$, by the Chapman-Kolmogorov equation, we have

$$p(s,\mathbf{x}\,|t,\mathbf{y}) = \int_{\mathbb{R}^n} p(s+\varepsilon,\mathbf{z}\,|t,\mathbf{y})\, p(s,\mathbf{x}\,|s+\varepsilon,\mathbf{z})\, d\mathbf{z} \tag{3.45}$$

Expanding $p(s+\varepsilon,\mathbf{z}\,|t,\mathbf{y})$ around $\mathbf{x}$, we have

$$p(s+\varepsilon,\mathbf{z}\,|t,\mathbf{y}) = p(s+\varepsilon,\mathbf{x}\,|t,\mathbf{y}) + \sum_{i=1}^{n} (z_i - x_i) \frac{\partial}{\partial x_i} p(s+\varepsilon,\mathbf{x}\,|t,\mathbf{y})$$

$$+ \frac{1}{2} \sum_{i,j=1}^{n} (z_i - x_i)(z_j - x_j) \frac{\partial^2}{\partial x_i \partial x_j} p(s+\varepsilon,\mathbf{x}\,|t,\mathbf{y}) + \cdots$$

Replacing this expression in (3.45) we get

$$p(s,\mathbf{x}\,|t,\mathbf{y}) = p(s+\varepsilon,\mathbf{x}\,|t,\mathbf{y}) + \sum_{i=1}^{n} \int_{\mathbb{R}^n} (z_i - x_i)\, p(s,\mathbf{x}\,|s+\varepsilon,\mathbf{z})\, d\mathbf{z}$$

$$\times \frac{\partial}{\partial x_i} p(s+\varepsilon,\mathbf{x}\,|t,\mathbf{y})$$

$$+ \frac{1}{2} \sum_{i,j=1}^{n} \int_{\mathbb{R}^n} (z_i - x_i)(z_j - x_j)\, p(s,\mathbf{x}\,|s+\varepsilon,\mathbf{z})\, d\mathbf{z}$$

$$\times \frac{\partial^2}{\partial x_i \partial x_j} p(s+\varepsilon,\mathbf{x}\,|t,\mathbf{y}) + \cdots$$

where we have applied that $\int_{\mathbb{R}^n} p(s,\mathbf{x}\,|s+\varepsilon,\mathbf{z})\, d\mathbf{z} = 1$.

[4] We present here a heuristic derivation of Kolmogorov's PDEs, based on Kuo (2006), Section 10.9 and Kallianpur and Sundar (2014), Section 8.3. For a rigorous proof, see, for example, Section 8.6 of the latter reference.

Reordering, dividing by ε, letting $\varepsilon \to 0^+$, and neglecting higher-order terms, we have

$$\begin{aligned}
&\lim_{\varepsilon\to 0^+} \frac{p(s,\mathbf{x}\,|t,\mathbf{y}) - p(s+\varepsilon,\mathbf{x}\,|t,\mathbf{y})}{\varepsilon} \\
&= \sum_{i=1}^{n} \lim_{\varepsilon\to 0^+} \frac{1}{\varepsilon} \int_{\mathbb{R}^n} (z_i - x_i)\, p(s,\mathbf{x}\,|s+\varepsilon,\mathbf{z})\, d\mathbf{z} \\
&\times \frac{\partial}{\partial x_i} p(s+\varepsilon,\mathbf{x}\,|t,\mathbf{y}) \\
&+ \frac{1}{2} \sum_{i,j=1}^{n} \lim_{\varepsilon\to 0^+} \frac{1}{\varepsilon} \int_{\mathbb{R}^n} (z_i - x_i)\left(z_j - x_j\right) p(s,\mathbf{x}\,|s+\varepsilon,\mathbf{z})\, d\mathbf{z} \\
&\times \frac{\partial^2}{\partial x_i \partial x_j} p(s+\varepsilon,\mathbf{x}\,|t,\mathbf{y})
\end{aligned}$$

Applying ii)' and iii)' of Lemma 3.57, we obtain expression (3.41). On the other hand

$$P(t,\mathbf{x}\,|t,B) = \mathbf{1}_B(\mathbf{x}) = \int_B p(t,\mathbf{x}\,|t,\mathbf{y})\, d\mathbf{y} \tag{3.46}$$

where we have applied Definition 3.38. From expression (3.46) we conclude that $p(t,\mathbf{x}\,|t,\mathbf{y}) = \delta(\mathbf{y}-\mathbf{x})$, which is expression (3.42).

For the forward PDE, again by Chapman-Kolmogorov, we have for $s < u < t$

$$p(s,\mathbf{x}\,|t,\mathbf{y}) = \int_{\mathbb{R}^n} p(u,\mathbf{z}\,|t,\mathbf{y})\, p(s,\mathbf{x}\,|u,\mathbf{z})\, d\mathbf{z}$$

Differentiating with respect to u, we get

$$0 = \int_{\mathbb{R}^n} \left[p(s,\mathbf{x}\,|u,\mathbf{z}) \frac{\partial}{\partial u} p(u,\mathbf{z}\,|t,\mathbf{y}) + p(u,\mathbf{z}\,|t,\mathbf{y}) \frac{\partial}{\partial u} p(s,\mathbf{x}\,|u,\mathbf{z}) \right] d\mathbf{z}$$

Replacing in this expression the value of $\frac{\partial}{\partial u} p(u,\mathbf{z}\,|t,\mathbf{y})$ taken from the backward Kolmogorov equation we obtain

$$\begin{aligned}
0 = \int_{\mathbb{R}^n} &-p(s,\mathbf{x}\,|u,\mathbf{z}) \left[\sum_{i=1}^{n} \rho_i(u,\mathbf{z}) \frac{\partial}{\partial z_i} p(u,\mathbf{z}\,|t,\mathbf{y}) \right. \\
&\left. + \frac{1}{2} \sum_{i,j=1}^{n} q_{ij}(u,\mathbf{z}) \frac{\partial^2}{\partial z_i \partial z_j} p(u,\mathbf{z}\,|t,\mathbf{y}) \right] \\
&+ p(u,\mathbf{z}\,|t,\mathbf{y}) \frac{\partial}{\partial u} p(s,\mathbf{x}\,|u,\mathbf{z})\, d\mathbf{z}
\end{aligned}$$

Integrating by parts the first two terms, one time the first term and two times the second term, and taking into account that p is a density function, we obtain

$$0 = \int_{\mathbb{R}^n} p(u, \mathbf{z} | t, \mathbf{y}) \left\{ \sum_{i=1}^{n} \frac{\partial}{\partial z_i} [p(s, \mathbf{x} | u, \mathbf{z}) \rho_i(u, \mathbf{z})] \right.$$
$$- \frac{1}{2} \sum_{i,j=1}^{n} \frac{\partial^2}{\partial z_i \partial z_j} \left[p(s, \mathbf{x} | u, \mathbf{z}) q_{ij}(u, \mathbf{z})\right]$$
$$\left. + \frac{\partial}{\partial u} p(s, \mathbf{x} | u, \mathbf{z}) \right\} d\mathbf{z}$$

Since the identity holds for all transition probability densities $p(u, \mathbf{z} | t, \mathbf{y})$, we obtain expression (3.43). Expression (3.44) is obtained similarly to (3.42), just by replacing t by s. □

Example 3.62. From Exercise 3.58 we know that one-dimensional Brownian motion is a diffusion with drift and diffusion coefficients $\rho(t, x) = 0$ and $q(t, x) = 1$. Then, its associated transition probability density, $p(s, x | t, y)$, satisfies the Kolmogorov forward PDE

$$\frac{\partial}{\partial t} p(s, x | t, y) - \frac{1}{2} \frac{\partial^2}{\partial y^2} p(s, x | t, y) = 0$$
$$p(s, x | s, y) = \delta(x - y)$$

whose solution is

$$p(s, x | t, y) = \frac{1}{\sqrt{2\pi(t-s)}} \exp\left\{ -\frac{1}{2} \frac{(x-y)^2}{t-s} \right\}$$

that was previously obtained in Exercise 3.43.

Feynman-Kac Formula

In this subsection we state another result relating PDEs and SDEs, the Feynman-Kac Theorem, which plays a key role in financial applications.

Theorem 3.63. ***(Feynman-Kac)*** *Consider the SDE*

$$d\mathbf{X}_t = \mathbf{b}(t, \mathbf{X}_t) dt + \mathbf{\Sigma}(t, \mathbf{X}_t) d\mathbf{W}_t, \quad 0 \le t \le T \tag{3.47}$$

satisfying the conditions of Theorem 3.59. Let u be a solution of the PDE

$$\frac{\partial u}{\partial t}(t, \mathbf{x}) + \sum_{i=1}^{n} \rho_i(t, \mathbf{x}) \frac{\partial u}{\partial x_i}(t, \mathbf{x})$$
$$+ \frac{1}{2} \sum_{i,j=1}^{n} q_{ij}(t, \mathbf{x}) \frac{\partial^2 u}{\partial x_i \partial x_j}(t, \mathbf{x}) - r(t, \mathbf{x}) u(t, \mathbf{x}) = 0, \quad t < T$$
$$u(T, \mathbf{x}) = \phi(\mathbf{x}) \tag{3.48}$$

where ϕ is a given function, r is a continuous function, and $\rho_i(t,\mathbf{x}) = b_i(t,\mathbf{x})$, $q_{ij}(t,\mathbf{x}) = \left[\boldsymbol{\Sigma}(t,\mathbf{x})\boldsymbol{\Sigma}^\top(t,\mathbf{x})\right]_{ij}$. Assume that the condition

$$\sum_{i=1}^n\sum_{j=1}^d \mathbb{E}\left[\int_0^T e^{-\int_0^t r(s,\mathbf{X}_s)ds}\left|\frac{\partial u}{\partial x_i}(t,\mathbf{X}_t)\boldsymbol{\Sigma}_{ij}(t,\mathbf{X}_t)\right|^2 dt\right] < \infty \tag{3.49}$$

holds. Then, the solution u can be represented as

$$u(s,\mathbf{x}) = \mathbb{E}\left[e^{-\int_s^T r(t,\mathbf{X}_t)dt}\phi(\mathbf{X}_T)\,|\mathbf{X}_s = \mathbf{x}\right]$$

where $\mathbf{X}$ solves the SDE (3.47) with $\mathbf{X}_s = \mathbf{x}$.

Proof. Define the process

$$g_t := e^{-\int_s^t r(u,\mathbf{X}_u)du}$$

which solves the differential equation $dg_t = -r(t,\mathbf{X}_t)\,g_t dt$, and consider the process $Y_t = g_t u(t,\mathbf{X}_t)$. Using integration by parts and multidimensional Itô's formula, we have

$$\begin{aligned}
dY_t &= g_t du(t,\mathbf{X}_t) + u(t,\mathbf{X}_t)\,dg_t = g_t\left[du(t,\mathbf{X}_t) - r(t,\mathbf{X}_t)\,u(t,\mathbf{X}_t)\,dt\right]\\
&= g_t\left\{\left[\frac{\partial u}{\partial t}(t,\mathbf{X}_t) + \sum_{i=1}^n \rho_i(t,\mathbf{X}_t)\frac{\partial u}{\partial x_i}(t,\mathbf{X}_t)\right.\right.\\
&\left.+\frac{1}{2}\sum_{i,j=1}^n q_{ij}(t,\mathbf{X}_t)\frac{\partial^2 u}{\partial x_i\partial x_j}(t,\mathbf{X}_t) - r(t,\mathbf{X}_t)\,u(t,\mathbf{X}_t)\right]dt\\
&\left.+\sum_{i=1}^n\sum_{j=1}^d \boldsymbol{\Sigma}_{ij}(t,\mathbf{X}_t)\frac{\partial u}{\partial x_i}(t,\mathbf{X}_t)\,dW_t^{(j)}\right\}
\end{aligned}$$

from where we get

$$\begin{aligned}
Y_T &= g_T u(T,\mathbf{X}_T)\\
&= u(s,\mathbf{X}_s) + \int_s^T g_t\left[\frac{\partial u}{\partial t}(t,\mathbf{X}_t) + \sum_{i=1}^n \rho_i(t,\mathbf{X}_t)\frac{\partial u}{\partial x_i}(t,\mathbf{X}_t)\right.\\
&\left.+\frac{1}{2}\sum_{i,j=1}^n q_{ij}(t,\mathbf{X}_t)\frac{\partial^2 u}{\partial x_i\partial x_j}(t,\mathbf{X}_t) - r(t,\mathbf{X}_t)\,u(t,\mathbf{X}_t)\right]dt\\
&+\int_s^T g_t\sum_{i=1}^n\sum_{j=1}^d \boldsymbol{\Sigma}_{ij}(t,\mathbf{X}_t)\frac{\partial u}{\partial x_i}(t,\mathbf{X}_t)\,dW_t^{(j)}
\end{aligned}$$

where we have used that $g_s = 1$. As u is solution of the PDE (3.48), the first integral is equal to zero, and from condition (3.49), the last integral has zero expectation. Thus, we conclude that

$$u(s,\mathbf{x}) = \mathbb{E}[Y_T \,|\mathbf{X}_s = \mathbf{x}] = \mathbb{E}\left[e^{-\int_s^T r(t,\mathbf{X}_t)dt} u(T,\mathbf{X}_T)\,|\mathbf{X}_s = \mathbf{x}\right]$$
$$= \mathbb{E}\left[e^{-\int_s^T r(t,\mathbf{X}_t)dt} \phi(\mathbf{X}_T)\,|\mathbf{X}_s = \mathbf{x}\right]$$

and then $u(T,\mathbf{x}) = \phi(\mathbf{x})$. □

Remark 3.64. We have assumed that $u(t,\mathbf{x})$ is a solution of equation (3.48). A set of sufficient conditions for the existence of a solution can be found in Karatzas and Shreve (1991), page 368.

Remark 3.65. It is possible to show that the solution $u(t,\mathbf{x})$ is unique in the set of functions belonging to $C^{1,2}(\mathbb{R}\times\mathbb{R}^n)$ which are bounded on $K\times\mathbb{R}^n$ for each compact $K\subset\mathbb{R}$ (Oksendal (1998), Theorem 8.2.1).

Exercise 3.66. Solve the PDE

$$\frac{\partial u}{\partial t}(t,x) + \frac{1}{2}\sigma^2\frac{\partial^2 u}{\partial x^2}(t,x) = 0$$
$$u(T,x) = x^2$$

3.8 Solutions to the Exercises

Exercise 3.9

a) In this case, we have $\mathbf{b}(t,\mathbf{x}) = b(x) = 3x^{1/3}$ and $\mathbf{\Sigma}(t,\mathbf{x}) = \Sigma(x) = 3x^{2/3}$. Both b and Σ are differentiable in $(0,+\infty)$ and have unbounded derivative, thus it does not exist $C\in\mathbb{R}$ such that

$$\frac{|b(x)-b(y)|}{|x-y|} + \frac{|\Sigma(x)-\Sigma(y)|}{|x-y|} \le C \quad \forall x,y\in(0,\infty)$$

and then, the Lipschitz condition is not satisfied.

b) Let us check that $X_t^{(a)}$ satisfies i)-iii) of Definition 3.1.

i) By its definition, $X_t^{(a)} = f(W_t)$, with $f(x) = (x-a)^3\,\mathbf{1}_{\{x\ge a\}}$. Since f is continuous, $X_t^{(a)}$ is also continuous . Moreover, as the initial value is deterministic, $\mathcal{F}_t^{\xi,W} = \mathcal{F}_t^W$, and $X_t^{(a)}$ is adapted with respect to $\mathcal{F}_t^{\xi,W}$.

ii) $\int_0^T \left(\left|b\left(X_t^{(a)}\right)\right| + \left|\Sigma\left(X_t^{(a)}\right)\right|^2\right) dt < \infty$ as the integral of a bounded and continuous function on a bounded interval.

iii) The function f is of class C^2, with derivatives $f'(x) = 3(x-a)^2 \mathbf{1}_{\{x\geq a\}}$ and $f''(x) = 6(x-a)\mathbf{1}_{\{x\geq a\}}$. Then, by Itô's formula, we get

$$\begin{aligned} X_t^{(a)} &= f(W_t) = f(W_0) + \frac{1}{2}\int_0^t 6(W_s - a)\mathbf{1}_{\{W_s\geq a\}}ds \\ &+ \int_0^t 3(W_s-a)^2 \mathbf{1}_{\{W_s\geq a\}}dW_s \\ &= 3\int_0^t X_s^{1/3}ds + 3\int_0^t X_s^{2/3}dW_s \end{aligned}$$

Therefore, we have proven that $X_t^{(a)}$ is a solution of the SDE (3.4) for every $a \geq 0$, and then, it has an infinite number of solutions.

Exercise 3.10

a) We have $\mathbf{b}(t,\mathbf{x}) = b(x) = x^3$ and $\mathbf{\Sigma}(t,\mathbf{x}) = \Sigma(x) = x^2$. Then

$$\frac{|b(x)| + |\Sigma(x)|}{1+|x|} = \frac{|x^3| + |x^2|}{1+|x|} \overset{x\to\infty}{\longrightarrow} \infty$$

and therefore, we cannot find a constant D such that the linear growth condition (3.3) is satisfied.

b) For the conditions in Definition (3.1), we have:

i) X_t is continuous in $[0, T_1)$ by its definition and, similarly to Exercise 3.9, X_t is adapted with respect to $\mathcal{F}_t^{\xi,W}$.

ii) Since $t < T_1$, continuity of Brownian paths implies that $1 - W_s$ is bounded away from zero on $[0,t]$. Therefore, condition ii) is satisfied.

iii) The function $f(x) = \frac{1}{1-x}$ is of class C^2 on $[0,1)$, with $f'(x) = \dfrac{1}{(1-x)^2}$ and $f''(x) = \dfrac{2}{(1-x)^3}$. Then, by Itô's formula

$$\begin{aligned} X_t &= f(W_t) = f(W_0) + \frac{1}{2}\int_0^t \frac{2}{(1-W_s)^3}ds + \int_0^t \frac{1}{(1-W_s)^2}dW_s \\ &= 1 + \int_0^t X_s^3 ds + \int_0^t X_s^2 dW_s \end{aligned}$$

Therefore, $X_t = \dfrac{1}{1-W_t}$ is a solution of (3.5) which is not a global solution since it is only valid on $[0, T_1)$.
In the present example, $\lim_{t\to T_1^-} X_t = +\infty$, and therefore the solution explodes at the random time T_1.[5]

[5] A solution of an SDE is said to explode if its magnitude becomes unbounded in finite time. The corresponding random time is called the explosion time. See, for example, Klebaner (2005), Section 6.6 and Karatzas and Shreve (1991), Section 5.5, for a more detailed analysis of explosions in SDEs.

Exercise 3.20

This is a scalar linear SDE of the type (3.12) with $d = 1$, $F = f = g_1 = 0$ and $G_1(t) = \sigma$. Then, the fundamental solution of expression (3.14) is

$$\Phi(t) = \exp\left\{-\frac{1}{2}\sigma^2 t + \sigma W_t\right\}$$

which replaced into equation (3.13) gives

$$X_t = X_0 \exp\left\{-\frac{1}{2}\sigma^2 t + \sigma W_t\right\}$$

Exercise 3.25

By its definition, Y_t is a Gaussian process. Its mean and covariance functions are

$$\mu(t) = \mathbb{E}[Y_t] = a(1-t) + bt + \mathbb{E}[W_t] - t\mathbb{E}[W_1] = a(1-t) + bt$$

where we have applied the martingale property of Brownian motion, and

$$\begin{aligned} c(s,t) &= \text{Cov}(Y_s, Y_t) = \text{Cov}(W_s - sW_1, W_t - tW_1) \\ &= \text{Cov}(W_s, W_t) - t\text{Cov}(W_s, W_1) - s\text{Cov}(W_1, W_t) + st\text{Cov}(W_s, W_1) \\ &= s \wedge t - ts - st + st = s \wedge t - st \end{aligned}$$

where we have used Theorem 1.15. As Gaussian processes are completely determined by their mean and covariance functions, X and Y have the same distribution.

Note that the process Y_t also satisfies $\lim_{t\to 1^-} Y_t = b$, and then, Y_t can be extended continuously to $[0, 1]$. Therefore, Y_t is a Brownian bridge from a to b, but it cannot be considered as a solution of equation (3.22) because the term with W_1 makes Y_t non-adapted with respect to $\mathcal{F}_t^W$.

Exercise 3.28

Equation (3.25) is not a linear SDE. We have that $\mu(x) = \sqrt{1+x^2} + \frac{1}{2}x$, $\sigma(x) = \sqrt{1+x^2}$, and then $\sigma'(x) = \dfrac{x}{\sqrt{1+x^2}}$. On the other hand

$$\varkappa(x) = \frac{\mu(x)}{\sigma(x)} - \frac{1}{2}\sigma'(x) = \frac{\sqrt{1+x^2} + \frac{1}{2}x}{\sqrt{1+x^2}} - \frac{x}{2\sqrt{1+x^2}} = 1$$

and therefore, $\varkappa'(x) = 0$. Thus, the SDE (3.25) satisfies the conditions in Lemma 3.27 with

$$Z_t = \lambda(X_t) = \int_0^{X_t} \frac{dy}{\sqrt{1+y^2}} = \ln\left(X_t + \sqrt{1+X_t^2}\right) \tag{3.50}$$

Then, Z_t satisfies the SDE

$$dZ_t = dt + dW_t$$

whose solution is

$$Z_t = Z_0 + t + W_t = t + W_t \tag{3.51}$$

where we have applied that $Z_0 = \lambda(X_0) = 0$.

Equating the right hand side of expressions (3.50) and (3.51) we obtain

$$t + W_t = \ln\left(X_t + \sqrt{1 + X_t^2}\right)$$

To solve equation (3.25) we need to find X_t as a function of W_t from this expression. Taking exponential, reordering terms and squaring both sides of the equation, we have

$$2X_t e^{(t+W_t)} = e^{2(t+W_t)} - 1$$

from where we finally obtain

$$X_t = \frac{1}{2}\left[e^{(t+W_t)} - e^{-(t+W_t)}\right] = \sinh(t + W_t)$$

Exercise 3.33

The coefficients are given by $\mu(t,x) = 2tx$ and $\sigma(t,x) = t^2 + x$. Let us check first the necessary condition (3.29). Taking derivatives we obtain

$$\frac{\partial \mu}{\partial x} = 2t$$

$$\frac{\partial \sigma}{\partial t} + \frac{1}{2}\frac{\partial^2 \sigma}{\partial x^2} = 2t$$

and then, the necessary condition holds. The SDE (3.30) is exact if we can obtain $f(t,x)$ from the system (3.27), (3.28), which in our case is

$$2tx = \frac{\partial f}{\partial t} + \frac{1}{2}\frac{\partial^2 f}{\partial x^2}$$

$$t^2 + x = \frac{\partial f}{\partial x}$$

Integrating the second equation with respect to x, we obtain

$$f(t,x) = \int \left(t^2 + x\right) dx + F(t) = t^2 x + \frac{x^2}{2} + F(t)$$

Differentiating and replacing the derivatives in the first equation, we get

$$2tx = 2tx + F'(t) + \frac{1}{2}$$

whose solution is

$$F(t) = -\frac{t}{2}$$

and then

$$f(t,x) = t^2 x + \frac{x^2}{2} - \frac{t}{2}$$

from where we have

$$X_t = f(t, W_t) + C = t^2 W_t + \frac{W_t^2}{2} - \frac{t}{2} + C$$

Applying that $X_0 = 0$, we finally obtain

$$X_t = t^2 W_t + \frac{W_t^2}{2} - \frac{t}{2}$$

Exercise 3.35

With the notation of Proposition 3.34, we have $f(t,x) = \frac{1}{x}$, $g(t) = \alpha$ and $\rho_t = \exp\left\{\frac{\alpha^2 t}{2} - \alpha W_t\right\}$. From equation (3.31) we have

$$dY_t = \rho_t \frac{1}{Y_t/\rho_t} dt = \frac{\rho_t^2}{Y_t} dt$$

Solving this ODE by separation of variables, we obtain

$$\frac{Y_t^2}{2} = \int_0^t \rho_s^2 ds + C = \int_0^t e^{\alpha^2 s - 2\alpha W_s} ds + C$$

Making $t = 0$ we obtain $C = \frac{Y_0^2}{2}$, and then

$$Y_t^2 = Y_0^2 + 2\int_0^t e^{\alpha^2 s - 2\alpha W_s} ds$$

Using $Y_t = \rho_t X_t$, $\rho_0 = 1$ and the expression of ρ_t, we finally obtain

$$X_t = \exp\left\{-\frac{\alpha^2 t}{2} + \alpha W_t\right\}\left[X_0^2 + 2\int_0^t e^{\alpha^2 s - 2\alpha W_s} ds\right]^{1/2}$$

Exercise 3.43

Writing $W_t = (W_t - W_s) + W_s$, and taking into account that $W_t - W_s \sim N(0, t-s)$, we have that the distribution of W_t conditioned to $W_s = x$ is $N(x, t-s)$, for $s < t$. Then

$$P(s,x\,|t,y) = \mathbb{P}[W_t \le y\,|W_s = x] = \frac{1}{\sqrt{2\pi(t-s)}}\int_{-\infty}^{y} e^{-\frac{(w-x)^2}{2(t-s)}} dw$$

On the other hand

$$\int_{-\infty}^{+\infty} P(u,z\,|t,y) P(s,x\,|u,dz)$$
$$= \frac{1}{2\pi\sqrt{(t-u)(u-s)}}\int_{-\infty}^{y}\left[\int_{-\infty}^{+\infty} e^{-\frac{(v-z)^2}{2(t-u)}} e^{-\frac{(z-x)^2}{2(u-s)}} dz\right] dv \tag{3.52}$$

Working apart with the exponent, we have

$$
\begin{aligned}
&-\frac{(v-z)^2}{2(t-u)}-\frac{(z-x)^2}{2(u-s)} \\
&\quad=-\frac{(t-s)z^2-2[(t-u)x+(u-s)v]z+(t-u)x^2+(u-s)v^2}{2(t-u)(u-s)} \\
&\quad=-\frac{1}{2}\frac{(t-s)}{(t-u)(u-s)}\left[z^2-2\frac{[(t-u)x+(u-s)v]}{t-s}z\right. \\
&\quad\left.+\frac{(t-u)x^2+(u-s)v^2}{t-s}\right] \\
&\quad=-\frac{1}{2}\frac{(t-s)}{(t-u)(u-s)} \\
&\quad\times\left[\left(z-\frac{[(t-u)x+(u-s)v]}{t-s}\right)^2-\frac{[(t-u)x+(u-s)v]^2}{(t-s)^2}\right. \\
&\quad\left.+\frac{(t-u)x^2+(u-s)v^2}{t-s}\right] \\
&\quad=-\frac{\left[z-\frac{[(t-u)x+(u-s)v]}{t-s}\right]^2}{\frac{2(t-u)(u-s)}{t-s}}-\frac{(v-x)^2}{2(t-s)}
\end{aligned}
$$

Replacing this result in (3.52) we get

$$
\begin{aligned}
&\int_{-\infty}^{+\infty} P(u,z\,|t,y)\,P(s,x\,|u,dz) \\
&\quad=\frac{1}{2\pi\sqrt{(t-u)(u-s)}} \\
&\quad\times\int_{-\infty}^{y} e^{-\frac{(v-x)^2}{2(t-s)}}\left[\int_{-\infty}^{+\infty}\exp\left\{-\frac{\left[z-\frac{[(t-u)x+(u-s)v]}{t-s}\right]^2}{\frac{2(t-u)(u-s)}{t-s}}\right\}dz\right]dv \\
&\quad=\frac{1}{\sqrt{2\pi(t-s)}}\int_{-\infty}^{y} e^{-\frac{(v-x)^2}{2(t-s)}}\,dv \\
&\quad=P(s,x\,|t,y)
\end{aligned}
$$

Exercise 3.50

The mean-reverting OU process is determined by expression (3.17):

$$dX_t=-\alpha(X_t-\mu)\,dt+\sigma dW_t$$

and then $\mathbf{b}(t,\mathbf{x})=-\alpha(x-\mu)$, $\boldsymbol{\Sigma}(t,\mathbf{x})=\sigma$. As X_t is continuous, by Corollary 3.49, the infinitesimal generator can be obtained from (3.36) as

$$(\mathcal{A}_t f)(x)=\frac{1}{2}\sigma^2 f''(x)-\alpha(x-\mu)f'(x)$$

Similarly, for the GBM we have (expression (3.20)):

$$dX_t = \mu X_t dt + \sigma X_t dW_t$$

Then $\mathbf{b}(t, \mathbf{x}) = \mu x$, $\mathbf{\Sigma}(t, \mathbf{x}) = \sigma x$, and

$$(\mathcal{A}_t f)(x) = \frac{1}{2}\sigma^2 x^2 f''(x) + \mu x f'(x)$$

Exercise 3.58

We already know, from Exercise 3.43, that the transition probability function of W_t is

$$P(s, x \,|t, y) = \frac{1}{\sqrt{2\pi(t-s)}} \int_{-\infty}^{y} e^{-\dfrac{(u-x)^2}{2(t-s)}} du$$

We also know that W_t is a Markov process with continuous paths, then we just need to check conditions i) of Definition 3.54 and ii)'-iii)' of Lemma 3.57.

i)

$$\begin{aligned}
\lim_{\varepsilon \to 0^+} \frac{1}{\varepsilon} \int_{\|\mathbf{y}-\mathbf{x}\| \geq c} P(t, \mathbf{x} \,|t+\varepsilon, d\mathbf{y}) &= \lim_{\varepsilon \to 0^+} \frac{1}{\varepsilon} \mathbb{P}\left[|W_{t+\varepsilon} - x| \geq c \,|W_t = x\right] \\
&= \lim_{\varepsilon \to 0^+} \frac{1}{\varepsilon} \mathbb{P}\left[|W_{t+\varepsilon} - x| \geq c\right] \\
&\leq \lim_{\varepsilon \to 0^+} \frac{1}{\varepsilon} \mathbb{P}\left[|W_\varepsilon| \geq c\right] \\
&\leq \lim_{\varepsilon \to 0^+} \frac{1}{\varepsilon} \frac{1}{c^4} \mathbb{E}\left[W_\varepsilon^4\right] \\
&= \lim_{\varepsilon \to 0^+} \frac{1}{\varepsilon} \frac{1}{c^4} 3\varepsilon^2 = 0
\end{aligned}$$

where we have used the generalized Chebyshev inequality and that $\mathbb{E}\left[W_t^4\right] = 3t^2$ (see Problem 1 of Chapter 1).[6]

ii)'

$$\begin{aligned}
\rho(t, x) &= \lim_{\varepsilon \to 0^+} \frac{1}{\varepsilon} \mathbb{E}\left[W_{t+\varepsilon} - W_t \,|W_t = x\right] \\
&= \lim_{\varepsilon \to 0^+} \frac{1}{\varepsilon} \frac{1}{\sqrt{2\pi\varepsilon}} \int_{-\infty}^{+\infty} (y-x)\, e^{-\dfrac{(y-x)^2}{2\varepsilon}} dy \\
&= \lim_{\varepsilon \to 0^+} \frac{1}{\varepsilon} \frac{1}{\sqrt{2\pi\varepsilon}} \int_{-\infty}^{+\infty} u e^{-\dfrac{u^2}{2\varepsilon}} du = 0
\end{aligned}$$

[6]The generalized Chebyshev inequality is given by the following result (see, Pozniak (2009), Theorem 4.1):

Theorem 3.67. *Let $g : \mathbb{R} \to \mathbb{R}$ be a non-negative non-decreasing function defined on $[0, \infty)$ and ξ be a random variable such that $\mathbb{E}\{g(|\xi|)\} < \infty$. Then, for any non-negative value $a \geq 0$, the following inequality holds*

$$\mathbb{E}\{g(|\xi|)\} \geq g(a)\, \mathbb{P}\{|\xi| \geq a\}$$

iii)'

$$
\begin{aligned}
q(t,x) &= \lim_{\varepsilon\to0^+}\frac{1}{\varepsilon}\mathbb{E}\left[(W_{t+\varepsilon}-W_t)^2\,|W_t=x\right] \\
&= \lim_{\varepsilon\to0^+}\frac{1}{\varepsilon}\frac{1}{\sqrt{2\pi\varepsilon}}\int_{-\infty}^{+\infty}(y-x)^2\,e^{-\frac{(y-x)^2}{2\varepsilon}}\,dy \\
&= \lim_{\varepsilon\to0^+}\frac{1}{\varepsilon}\frac{1}{\sqrt{2\pi\varepsilon}}\int_{-\infty}^{+\infty}u^2e^{-\frac{u^2}{2\varepsilon}}\,dy \\
&= \lim_{\varepsilon\to0^+}\frac{1}{\varepsilon}\varepsilon = 1
\end{aligned}
$$

Thus, one-dimensional Brownian motion is a diffusion with drift and difusion coefficients $\rho(t,x)=0$ and $q(t,x)=1$.

Exercise 3.66

Assuming that condition (3.49) holds, if $u(s,x)$ is a solution of the PDE, then, by Theorem 3.48 we have

$$u(s,x)=\mathbb{E}\left[X_T^2\,|X_s=x\right]$$

where X_t satisfies the SDE $dX_t=\sigma dW_t$. To obtain the expectation, we need the distribution of $X_T\,|X_s=x$. Writing the SDE in integral form we obtain

$$X_T=X_s+\sigma\int_s^T dW_t=X_s+\sigma\,(W_T-W_s)$$

and then

$$X_T\,|X_s=x\sim N\left(x,\sigma^2\,(T-s)\right)$$

Therefore

$$
\begin{aligned}
u(s,x) &= \mathbb{E}\left[X_T^2\,|X_s=x\right]=\operatorname{var}\left[X_T\,|X_s=x\right]+\left(\mathbb{E}\left[X_T\,|X_s=x\right]\right)^2 \\
&= \sigma^2\,(T-s)+x^2
\end{aligned}
$$

3.9 Problems

1. Consider the process

$$X_t=\int_a^t f(s)\,dWs,\quad a\le t\le b$$

with $f\in L^2[a,b]$. Show that the transition probability function $P(s,x\,|t,y)=\mathbb{P}(X_t\le y\,|X_s=x)$, $a\le s<t\le b$, satisfies the Chapman-Kolmogorov equation, and conclude that X_t is a Markov process.

2. Let X_t be the mean-reverting Ornstein-Uhlenbeck process of equation (3.18).

a) Prove that X_t can be expressed, for $0 \leq s \leq t$ as

$$X_t = \mu + (X_s - \mu)\, e^{-\alpha(t-s)} + \sigma \int_s^t e^{-\alpha(t-u)} dW_u$$

b) Use the expression obtained in a) to find the transition probability function of X_t.

3. Let μ, σ be two non-zero constants and consider the SDE

$$dX_t = \mu dt + \sigma dW_t \tag{3.53}$$

a) Obtain the transition probability density, $p(s, x\,|t, y)$, associated to X_t.

b) Write down the corresponding Kolmogorov forward equation and verify that $p(s, x\,|t, y)$ satisfies it.

4. Let X_t be the geometric Brownian motion of equation (3.21), with $X_0 > 0$ and deterministic.

a) Obtain $\mathbb{E}[X_T\,|\mathcal{F}_t]$ for $t \leq T$.

b) Find a process g_s and a constant x such that

$$X_T = x + \int_0^T g_s dW_s$$

c) Find the transition probability function and the transition probability density of X_t.

d) Write down the associated Kolmogorov forward PDE and verify that the transition probability density found in c) satisfies the PDE.

5. Solve the PDE

$$\frac{\partial u}{\partial t}(t,x) + \mu x \frac{\partial u}{\partial x}(t,x) + \frac{1}{2}\sigma^2 x^2 \frac{\partial^2 u}{\partial x^2}(t,x) = 0, \quad t < T$$

$$u(T,x) = \ln\left(x^2\right)$$

3.10 Notes

In the second section, devoted to the existence and uniqueness of solutions, we have started directly with the multidimensional case with a view to presenting later the relationship between SDEs and PDEs. The definition of a solution of an SDE is taken from Kallianpur and Sundar (2014) and the theorem of existence and uniqueness is from Oksendal (1998). Other useful references have been Karatzas and Shreve (1991), Klebaner (2005), Kuo (2006) and Calin (2015).

The study of linear SDEs in Sections 3.2 and 3.3 is based primarily on Mao (2008). Other useful references are Karatzas and Shreve (1991), Oksendal (1998), Evans (2013) and Calin (2015).

The integration techniques of Section 3.4 are inspired by Calin (2015), except the part dedicated to transformations into linear SDEs, which is adapted from Schilling and Partzsch (2012).

Section 3.5 is mainly based on Kuo (2006), Kallianpur and Sundar (2014) and Oksendal (1998). The results regarding the infinitesimal generator and Dynkin's formula have been obtained, following Kallianpur and Sundar (2014), under the assumption that the functions belong to C_b^2. Similar results can be obtained for functions in C_0^2, the set of twice continuously differentiable functions with compact support (Oksendal, 1998).

Various definitions of diffusion process can be found in the academic literature (see, for example, Kuo (2006), Karatzas and Shreve (1991) or Protter (2005)). The definition we have used here is the one presented in Kuo (2006), Kallianpur and Sundar (2014), and Kloeden and Platen (1992). Similarly, different approaches to the study of this type of processes can be found (Karatzas and Shreve, 1991; Kuo, 2006): the semigroup approach, based on transition operators and infinitesimal generators, the partial differential equations approach, based on Kolmogorov's equations, and the stochastic differential equations approach, based on SDEs. We have taken an eclectic approach, presenting the essentials of the three approaches and the relationships between the different concepts, based mainly on Kuo (2006) and Kallianpur and Sundar (2014).

Although Kolmogorov's equations can be studied for general diffusion processes, we have carried out a study of these equations only for the case of diffusions obtained as solutions of SDEs, similarly to the treatment in Kallianpur and Sundar (2014), which is sufficient for our purposes.

The result known as *Feynman-Kac Theorem* traces back to the independent works of Feynman (1948) and Kac (1949) who obtained probabilistic representations of solutions to the Schrödinger equation. Our presentation of the Feynman-Kac formula is based on Kallianpur and Sundar (2014) and Oksendal (1998), and covers most financial applications. Other useful references are Kuo (2006) and Klebaner (2005). For a more general version of the theorem, including a source term, see Karatzas and Shreve (1991).

Chapter 4

Solutions to the Problems

Chapter 1

1. The moment generating function of W_t is

$$M_t(u) = \mathbb{E}\left[e^{uW_t}\right] = \mathbb{E}\left[e^{i(-i)W_t}\right] = \varphi_{W_t}(-iu) = \exp\left[-\frac{1}{2}(-iu)^2 t\right]$$
$$= \exp\left[\frac{1}{2}u^2 t\right]$$

where we have applied that the characteristic function of Brownian motion is $\varphi_{W_t}(u) = \exp\left[-\frac{1}{2}u^2 t\right]$. It is not difficult to prove by induction that the derivatives of $M_t(u)$ satisfy the recursive relation

$$M_t^{(n)}(u) = (n-1)\, t M_t^{(n-2)}(u) + ut M_t^{(n-1)}(u)$$

On the other hand, by the properties of moment generating functions, we know that $\mathbb{E}\left[W_t^n\right] = M_t^{(n)}(0)$. Therefore, we can write

$$\begin{aligned}
\mathbb{E}\left[W_t^{2k}\right] &= M_t^{(2k)}(0) = (2k-1)\, t M_t^{(2k-2)}(0) \\
&= (2k-1)\, t\, (2k-3)\, t M_t^{(2k-4)}(0) \\
&= (2k-1) \cdot (2k-3) \cdots 3 \cdot 1 t^k M_t(0) \\
&= (2k-1)!! t^k
\end{aligned}$$

2. The process X_t is obviously adapted. For the integrability, we know from equation (1.4) that $\mathbb{E}\left[|W_t|\right] < \infty$. On the other hand

$$\begin{aligned}
\mathbb{E}\left[\left|W_t^3\right|\right] &= \frac{1}{\sqrt{2\pi t}} \int_{-\infty}^{+\infty} \left|x^3\right| e^{-x^2/2t} dx = \frac{2}{\sqrt{2\pi t}} \int_0^{+\infty} x^3 e^{-x^2/2t} dx \\
&= \frac{4t^2}{\sqrt{2\pi t}} \int_0^{\infty} u e^{-u} du = \frac{4t^2}{\sqrt{2\pi t}} < \infty \qquad (4.1)
\end{aligned}$$

where we have applied that the first integrand is an even function. Then, $\mathbb{E}[|X_t|] < \infty$.

Finally, for $s < t$

$$\begin{aligned}
\mathbb{E}[X_t | \mathcal{F}_s] &= \mathbb{E}\left[aW_t^3 - tW_t | \mathcal{F}_s\right] \\
&= \mathbb{E}\left[a[(W_t - W_s) + W_s]^3 - t(W_t - W_s + W_s) | \mathcal{F}_s\right] \\
&= 3aW_s\mathbb{E}\left[(W_t - W_s)^2\right] + aW_s^3 - tW_s \\
&= aW_s^3 - [t - 3a(t-s)]W_s
\end{aligned}$$

where we have applied that Brownian motion increments are independent from the past, and that $W_t - W_s \sim N(0, t-s)$, which implies

$$\mathbb{E}[W_t - W_s] = \mathbb{E}\left[(W_t - W_s)^3\right] = 0$$

and $\mathbb{E}\left[(W_t - W_s)^2\right] = t - s$. Therefore, X_t is a martingale if, and only if, $t - 3a(t-s) = s$, i.e., $a = \frac{1}{3}$.

3. Both processes are clearly adapted. For the integrability of both processes, it suffices to use the martingale property of W_t, expression (4.1), and that

$$\mathbb{E}\left[\left|\int_0^t W_s ds\right|\right] \leq \mathbb{E}\left[\int_0^t |W_s| ds\right] = \int_0^t \mathbb{E}|W_s| ds < \infty$$

For the conditional expectations, we have

$$\begin{aligned}
\mathbb{E}[V_t | \mathcal{F}_s] &= t\mathbb{E}[W_t | \mathcal{F}_s] - \mathbb{E}\left[\int_0^t W_u du | \mathcal{F}_s\right] \\
&= tW_s - \int_0^s W_u du - \int_s^t \mathbb{E}[W_u | \mathcal{F}_s] du \\
&= tW_s - \int_0^s W_u du - \int_s^t W_s du \\
&= tW_s - \int_0^s W_u du - (t-s)W_s \\
&= sW_s - \int_0^s W_u du = V_s
\end{aligned}$$

and

$$\begin{aligned}
\mathbb{E}[X_t | \mathcal{F}_s] &= \mathbb{E}\left[W_t^3 | \mathcal{F}_s\right] - 3\left[\int_0^s W_u du + (t-s)W_s\right] \\
&= 3(t-s)W_s + W_s^3 - 3\int_0^s W_u du - 3(t-s)W_s \\
&= W_s^3 - 3\int_0^s W_u du = X_s
\end{aligned}$$

where, for the second process, we have used the computation carried out in Problem 2.

4. The process is clearly adapted. Moreover, by the triangle inequality, we have

$$\mathbb{E}\left[|X_t|\right] \le \frac{1}{d}\mathbb{E}\|\mathbf{W}_t\|^2 + t = \frac{1}{d}\sum_{j=1}^{d}\mathbb{E}\left[\left(W_t^{(j)}\right)^2\right] + t = 2t < \infty$$

On the other hand, using the properties of one-dimensional Brownian motion, we have, for $s \le t$

$$\begin{aligned}
\mathbb{E}\left[X_t \,|\mathcal{F}_s\right] &= \mathbb{E}\left[\frac{1}{d}\|\mathbf{W}_t\|^2 - t \,|\mathcal{F}_s\right] = \frac{1}{d}\mathbb{E}\left[\|\mathbf{W}_t\|^2 \,|\mathcal{F}_s\right] - t \\
&= \frac{1}{d}\sum_{j=1}^{d}\mathbb{E}\left[\left(W_t^{(j)}\right)^2 |\mathcal{F}_s\right] - t \\
&= \frac{1}{d}\sum_{j=1}^{d}\mathbb{E}\left\{\left[\left(W_t^{(j)} - W_s^{(j)}\right) + W_s^{(j)}\right]^2 |\mathcal{F}_s\right\} - t \\
&= \frac{1}{d}\sum_{j=1}^{d}\mathbb{E}\left\{\left[\left(W_t^{(j)} - W_s^{(j)}\right)^2\right] + \left(W_s^{(j)}\right)^2\right. \\
&\quad \left. +2W_s^{(j)}\left(W_t^{(j)} - W_s^{(j)}\right) |\mathcal{F}_s\right\} - t \\
&= \frac{1}{d}\sum_{j=1}^{d}\left\{(t-s) + \left(W_s^{(j)}\right)^2\right\} - t \\
&= \frac{1}{d}\left[d\,(t-s) + \|\mathbf{W}_s\|^2\right] - t \\
&= \frac{1}{d}\|\mathbf{W}_s\|^2 - s = X_s
\end{aligned}$$

5. The processes R_t and I_t are adapted by their definition. We first compute the conditional expectation of M_t. For $s < t$ we have

$$\begin{aligned}
\mathbb{E}\left[M_t \,|\mathcal{F}_s\right] &= \mathbb{E}\left\{\exp\left(i\left\langle \mathbf{u}, \mathbf{W}_t\right\rangle + \frac{t}{2}\|\mathbf{u}\|^2\right) |\mathcal{F}_s\right\} \\
&= \mathbb{E}\left\{\exp\left(i\left\langle \mathbf{u}, (\mathbf{W}_t - \mathbf{W}_s) + \mathbf{W}_s\right\rangle + \left(\frac{t-s}{2} + \frac{s}{2}\right)\|\mathbf{u}\|^2\right) |\mathcal{F}_s\right\} \\
&= M_s\mathbb{E}\left\{\exp\left(i\left\langle \mathbf{u}, \mathbf{W}_t - \mathbf{W}_s\right\rangle + \frac{t-s}{2}\|\mathbf{u}\|^2\right)\right\} \\
&= M_s\mathbb{E}\left\{\exp\left(i\left\langle \mathbf{u}, \mathbf{W}_t - \mathbf{W}_s\right\rangle\right)\right\}\exp\left(\frac{t-s}{2}\|\mathbf{u}\|^2\right) \\
&= M_s\exp\left(-\frac{t-s}{2}\|\mathbf{u}\|^2\right)\exp\left(\frac{t-s}{2}\|\mathbf{u}\|^2\right) = M_s
\end{aligned}$$

where we have used the properties and the characteristic function of d-dimensional Brownian motion.

Using the previous calculation, we obtain the conditional expectation of R_t as

$$\mathbb{E}[R_t |\mathcal{F}_s] = \mathbb{E}[\mathrm{Re}(M_t)|\mathcal{F}_s] = \mathrm{Re}\{\mathbb{E}[M_t|\mathcal{F}_s]\} = \mathrm{Re}(M_s) = R_s$$

For the integrability condition, we get

$$\mathbb{E}[|R_t|] \leq \mathbb{E}[|M_t|] = e^{\frac{1}{2}\|\mathbf{u}\|^2 t} < \infty$$

where we have applied that $|\exp(i\langle\mathbf{u},\mathbf{W}_t\rangle)| = 1$.

The proof for I_t is analogous.

Chapter 2

1. Since f is of class C^1, it is of finite variation, and the integration by parts formula yields

$$\begin{aligned} d[f(t)g(W_t)] &= f(t)\,dg(W_t) + g(W_t)\,df(t) \\ &= f(t)\left[\frac{1}{2}g''(W_t)\,dt + g'(W_t)\,dW_t\right] + g(W_t)f'(t)\,dt \\ &= \left[g(W_t)f'(t) + \frac{1}{2}f(t)g''(W_t)\right]dt + f(t)g'(W_t)\,dW_t \end{aligned}$$

where we have applied Itô's formula. Writing the previous expression in integral form, we have

$$\begin{aligned} [f(s)g(W_s)]_0^t = &\int_0^t f'(s)g(W_s)\,ds + \frac{1}{2}\int_0^t f(s)g''(W_s)\,ds \\ &+ \int_0^t f(s)g'(W_s)\,dW_s \end{aligned}$$

and rearranging the terms we obtain the desired result.

2. Taking $f(x) = x^{2k}$ we have that $W_t^{2k} = f(W_t)$, and

$$f'(x) = 2kx^{2k-1}, \quad f''(x) = 2k(2k-1)x^{2k-2}$$

Then, by Itô's formula

$$W_t^{2k} = W_0^{2k} + k(2k-1)\int_0^t W_s^{2k-2}ds + 2k\int_0^t W_s^{2k-1}dW_s$$

Taking expectations and applying the martingale property of stochastic integrals, we have

$$
\begin{aligned}
\mathbb{E}\left[W_t^{2k}\right] &= k\,(2k-1)\int_0^t \mathbb{E}\left[W_{t_1}^{2k-2}\right]dt_1 = k\,(2k-1)\int_0^t \mathbb{E}\left[W_{t_1}^{2(k-1)}\right]dt_1 \\
&= k\,(2k-1)\int_0^t (k-1)\,(2k-3)\left[\int_0^{t_1}\mathbb{E}\left[W_{t_2}^{2(k-2)}\right]dt_2\right]dt_1 \\
&= k\,(2k-1)\,(k-1)\,(2k-3)\int_0^t\int_0^{t_1}\mathbb{E}\left[W_{t_2}^{2(k-2)}\right]dt_2dt_1 = \cdots \\
&= k\,(k-1)\cdots 1\cdot(2k-1)\,(2k-3)\cdots 1 \\
&\times\int_0^t\int_0^{t_1}\cdots\int_0^{t_{k-1}}\mathbb{E}\left[W_{t_k}^0\right]dt_k\cdots dt_1 \\
&= k!\,(2k-1)!!\frac{t^k}{k!} \\
&= (2k-1)!!t^k
\end{aligned}
$$

which is the same result obtained in Problem 1 of Chapter 1 by using the moment generating function.

3. Using the definition of variance, the martingale property of stochastic integrals and Itô isometry, we have

$$
\begin{aligned}
\mathrm{Var}\left[\int_0^t e^{W_s}dW_s\right] &= \mathbb{E}\left\{\left[\int_0^t e^{W_s}dW_s\right]^2\right\} - \left\{\mathbb{E}\left[\int_0^t e^{W_s}dW_s\right]\right\}^2 \\
&= \mathbb{E}\left\{\left[\int_0^t e^{W_s}dW_s\right]^2\right\} = \mathbb{E}\left[\int_0^t e^{2W_s}ds\right] \\
&= \int_0^t \mathbb{E}\left[e^{2W_s}\right]ds \qquad (4.2)
\end{aligned}
$$

To evaluate the expectation, we use Itô's formula, with $f(x) = e^{2x}$, obtaining

$$
e^{2W_t} = 1 + 2\int_0^t e^{2W_s}ds + 2\int_0^t e^{2W_s}dW_s
$$

and

$$
\mathbb{E}\left[e^{2W_t}\right] = 1 + 2\int_0^t \mathbb{E}\left[e^{2W_s}\right]ds
$$

Making $f(t) := \mathbb{E}\left[e^{2W_t}\right]$ and differentiating, we obtain the differential equation

$$
f'(t) = 2f(t)
$$

whose solution with $f(0) = 1$ is

$$
f(t) = e^{2t}
$$

Replacing this result in (4.2) we finally obtain

$$
\mathrm{Var}\left[\int_0^t e^{W_s}dW_s\right] = \int_0^t e^{2s}ds = \frac{1}{2}\left(e^{2t} - 1\right)
$$

4. Taking $f(t,x) = e^{t/2}\sin x$ we have $M_t = f(t, W_t)$. Differentiating we obtain

$$\frac{\partial f}{\partial t}(t,x) = \frac{1}{2}e^{t/2}\sin x; \quad \frac{\partial f}{\partial x}(t,x) = e^{t/2}\cos x; \quad \frac{\partial^2 f}{\partial x^2}(t,x) = -e^{t/2}\sin x$$

Therefore

$$\frac{\partial f}{\partial t}(t,x) + \frac{1}{2}\frac{\partial^2 f}{\partial x^2}(t,x) = 0$$

and by Proposition 2.47 M_t is a local martingale.

On the other hand

$$\mathbb{E}\left\{\int_0^T \left[\frac{\partial f}{\partial x}(t, W_t)\right]^2 dt\right\} = \mathbb{E}\left[\int_0^T e^t \cos^2 W_t dt\right] = \int_0^T e^t \mathbb{E}\left[\cos^2 W_t\right] dt \tag{4.3}$$

In order to evaluate $\mathbb{E}\left[\cos^2 W_t\right]$, we apply Itô's formula to $g(x) = \cos^2 x$, with derivatives

$$g'(x) = -2\sin x \cos x = -\sin 2x; \quad g''(x) = -2\cos 2x$$

obtaining

$$\cos^2 W_t = 1 - \int_0^t \cos(2W_s)\, ds - \int_0^t \sin(2W_s)\, dW_s$$

Taking expectations we have

$$\begin{aligned}\mathbb{E}\left[\cos^2 W_t\right] &= 1 - \int_0^t \mathbb{E}\left[\cos(2W_s)\right] ds \\ &= 1 - 2\int_0^t \mathbb{E}\left[\cos^2 W_s\right] ds + t\end{aligned}$$

where we have used the trigonometric identity $\cos(2x) = 2\cos^2 x - 1$. Now, setting $h(t) = \mathbb{E}\left[\cos^2 W_t\right]$ and differentiating, we obtain

$$h'(t) + 2h(t) = 1$$

which is a first order linear ordinary differential equation with solution given by

$$h(t) = \frac{1}{2} + Ce^{-2t}$$

Taking into account that $h(0) = \mathbb{E}\left[\cos^2 0\right] = 1$, we obtain $C = \frac{1}{2}$, and then

$$h(t) = \mathbb{E}\left[\cos^2 W_t\right] = \frac{1 + e^{-2t}}{2}$$

Replacing this result in (4.3) we have

$$\mathbb{E}\left\{\int_0^T \left[\frac{\partial f}{\partial x}(t, W_t)\right]^2 dt\right\} = \int_0^T e^t \frac{1 + e^{-2t}}{2} dt = \int_0^T \cosh t\, dt = \sinh T < \infty$$

Therefore, again by Proposition 2.47, for every $T > 0$, M_t is a martingale for $0 \le t \le T$.

5. Taking $f(t,x) = \Phi\left(\frac{x}{\sqrt{T-t}}\right)$ we obtain the partial derivatives

$$\frac{\partial f}{\partial t}(t,x) = \frac{x}{2(T-t)\sqrt{T-t}}\phi\left(\frac{x}{\sqrt{T-t}}\right)$$
$$\frac{\partial f}{\partial x}(t,x) = \frac{1}{\sqrt{T-t}}\phi\left(\frac{x}{\sqrt{T-t}}\right)$$
$$\frac{\partial^2 f}{\partial x^2}(t,x) = -\frac{x}{(T-t)\sqrt{T-t}}\phi\left(\frac{x}{\sqrt{T-t}}\right)$$

where ϕ is the density function of a standard Gaussian random variable, and we have used the identitities $\Phi'(x) = \phi(x)$ and $\phi'(x) = -x\phi(x)$.

Taking into account the previous calculations we have that

$$\frac{\partial f}{\partial t}(t,x) + \frac{1}{2}\frac{\partial^2 f}{\partial x^2}(t,x) = 0$$

and by Proposition 2.47, $\Phi\left(\frac{W_t}{\sqrt{T-t}}\right)$ is a local martingale.

To prove that it is a martingale, we need to evaluate

$$\begin{aligned}
\mathbb{E}\left\{\int_0^T \left[\frac{\partial f}{\partial x}(t,W_t)\right]^2 dt\right\} &= \mathbb{E}\left\{\int_0^T \left[\frac{1}{\sqrt{T-t}}\phi\left(\frac{W_t}{\sqrt{T-t}}\right)\right]^2 dt\right\} \\
&= \int_0^T \frac{1}{T-t}\mathbb{E}\left[\phi^2\left(\frac{W_t}{\sqrt{T-t}}\right)\right] dt \\
&= \frac{1}{2\pi}\int_0^T \frac{1}{T-t}\mathbb{E}\left[e^{-\frac{W_t^2}{T-t}}\right] dt \qquad (4.4)
\end{aligned}$$

In this case, it can be seen that applying Itô's formula does not yield a differential equation for the expectation of the right-hand side. Therefore, we will take an alternative approach.

As $W_t \sim N(0,t)$, if we define the random variable $Z_t := \frac{1}{\sqrt{t}}W_t$, we have that $Z_t \sim N(0,1)$. Then, we can write

$$\begin{aligned}
\mathbb{E}\left[e^{-\frac{W_t^2}{T-t}}\right] &= \mathbb{E}\left[e^{-\frac{t}{T-t}Z_t^2}\right] = \frac{1}{\sqrt{2\pi}}\int_{-\infty}^{+\infty} e^{-\frac{t}{T-t}x^2}e^{-\frac{x^2}{2}}dx \\
&= \frac{1}{\sqrt{2\pi}}\sqrt{\frac{T-t}{T+t}}\int_{-\infty}^{+\infty} e^{-\frac{u^2}{2}}du = \sqrt{\frac{T-t}{T+t}}
\end{aligned}$$

Replacing this result in (4.4) we can write

$$\begin{aligned}
\mathbb{E}\left\{\int_0^T \left[\frac{\partial f}{\partial x}(t,W_t)\right]^2 dt\right\} &= \frac{1}{2\pi}\int_0^T \frac{1}{T-t}\sqrt{\frac{T-t}{T+t}}dt \\
&= \frac{1}{2\pi}\int_0^T \frac{1}{\sqrt{T^2-t^2}}dt \\
&= \frac{1}{2\pi}\left[\arcsin\left(\frac{t}{T}\right)\right]_0^T = \frac{1}{4} < \infty
\end{aligned}$$

and then, again by Proposition 2.47, $\Phi\left(\frac{W_t}{\sqrt{T-t}}\right)$ is a martingale on $[0,T]$.

6. a) Taking partial derivative with respect to t in both sides of

$$e^{\lambda x-\lambda^2 t/2}=\sum_{n=0}^{\infty} h_n(t,x)\,\lambda^n$$

we have

$$\begin{aligned}\frac{\partial}{\partial t}\left(e^{\lambda x-\lambda^2 t/2}\right)&=-\frac{\lambda^2}{2}e^{\lambda x-\lambda^2 t/2}=-\frac{\lambda^2}{2}\sum_{n=0}^{\infty} h_n(t,x)\,\lambda^n\\&=-\frac{1}{2}\sum_{n=2}^{\infty} h_{n-2}(t,x)\,\lambda^n\end{aligned}$$

and

$$\frac{\partial}{\partial t}\sum_{n=0}^{\infty} h_n(t,x)\,\lambda^n=\sum_{n=0}^{\infty}\frac{\partial h_n(t,x)}{\partial t}\lambda^n$$

Equating both results we obtain

$$\frac{\partial h_n}{\partial t}(t,x)=\begin{cases}0 & n=0,1\\ -\frac{1}{2}h_{n-2}(t,x) & n=2,3,\ldots\end{cases}\tag{4.5}$$

In a similar manner, it is possible to find the relations

$$\frac{\partial h_n}{\partial x}(t,x)=\begin{cases}0 & n=0\\ h_{n-1}(t,x) & n=1,2,\ldots\end{cases}\tag{4.6}$$

and

$$\frac{\partial^2 h_n}{\partial x^2}(t,x)=\begin{cases}0 & n=0,1\\ h_{n-2}(t,x) & n=2,3,\ldots\end{cases}\tag{4.7}$$

From equations (4.5) and (4.7), we obtain

$$\frac{\partial h_n}{\partial t}(t,W_t)+\frac{1}{2}\frac{\partial^2 h_n}{\partial x^2}(t,W_t)=0$$

and by Proposition 2.47, $h_n(t,W_t)$ is a local martingale.
Moreover, for any $T>0$

$$\mathbb{E}\left[\int_0^T\left(\frac{\partial h_n}{\partial x}(t,W_t)\right)^2 dt\right]=\mathbb{E}\left[\int_0^T (h_{n-1}(t,W_t))^2\,dt\right]<\infty$$

since $\mathbb{E}\left[(h_{n-1}(t,W_t))^2\right]$ is a polynomial on t (see Problem 1 of Chapter 1). Applying again Proposition 2.47, we conclude that $h_n(t,W_t)$ is a martingale on $[0,T]$.

b) Taking into account expressions (4.5)-(4.7) and applying Itô's formula we have

$$\begin{aligned} dh_{n+1}(t, W_t) &= \left[-\frac{1}{2}h_{n-1}(t, W_t) + \frac{1}{2}h_{n-1}(t, W_t)\right] dt + h_n(t, W_t)\, dW_t \\ &= h_n(t, W_t)\, dW_t \end{aligned}$$

7. a) By the hint, we know that $\|\mathbf{W}_t\| > 0$ for every $t > 0$ a.s. Hence, the processes $\frac{W_{it}}{\|\mathbf{W}_t\|}$, $i = 1, \ldots, n$, are progressively measurable. Moreover,

$$\mathbb{E}\left[\int_0^T \frac{W_{is}^2}{\|\mathbf{W}_s\|^2} ds\right] \leq \mathbb{E}\left[\int_0^T \frac{\|\mathbf{W}_s\|^2}{\|\mathbf{W}_s\|^2} ds\right] = T < \infty$$

Consequently, each stochastic integral is well defined and is a continuous martingale. Therefore, $\widetilde{W}_t$ is also a continuous martingale. On the other hand, by the properties of quadratic variation we can write

$$\begin{aligned} \left[\widetilde{W}_\cdot, \widetilde{W}_\cdot\right]_t &= \sum_{i=1}^n \sum_{j=1}^n \int_0^t \frac{W_{is}}{\|\mathbf{W}_s\|} \frac{W_{js}}{\|\mathbf{W}_s\|} d\left[W_{i\cdot}, W_{j\cdot}\right]_s \\ &= \sum_{i=1}^n \int_0^t \frac{W_{is}^2}{\|\mathbf{W}_s\|^2} ds = \int_0^t ds = t \end{aligned}$$

where we have applied that $\mathbf{W}_t$ is a n-dimensional Brownian motion, and then $\left[W_{i\cdot}, W_{j\cdot}\right]_s = \delta_{ij} s$. Finally, by Lévy's characterization theorem, $\widetilde{W}$ is a Brownian motion.

b) As $Z_t = \|\mathbf{W}_t\|^2 = W_{1t}^2 + \cdots + W_{nt}^2$, we have that $Z_t = f(W_{1t}, \ldots, W_{nt})$, with $f(x_1, \ldots, x_n) = x_1^2 + \cdots + x_n^2$. Then

$$\frac{\partial f}{\partial t} = 0, \quad \frac{\partial f}{\partial x_i} = 2x_i, \quad \frac{\partial^2 f}{\partial x_i \partial x_j} = 2\delta_{ij}$$

and applying the multidimensional Itô's formula we obtain

$$\begin{aligned} dZ_t &= \frac{1}{2} \cdot 2 \sum_{i=1}^n \sum_{j=1}^n \delta_{ij} d\left[W_{i\cdot}, W_{j\cdot}\right]_t + 2 \sum_{i=1}^n W_{it} dW_{it} \\ &= n dt + 2\|\mathbf{W}_t\| \sum_{i=1}^n \frac{W_{it}}{\|\mathbf{W}_t\|} dW_{it} \\ &= n dt + 2\sqrt{Z_t}\, d\widetilde{W}_t \end{aligned}$$

8. The problem asks us to express the random variables in the form $Y = \mathbb{E}[Y] + \int_0^T g_s dW_s$. Instead of using directly Theorem 2.85, it is easiest to work with martingale representation as follows.

   Let $M_t = \mathbb{E}[Y \,|\mathcal{F}_t]$. By Corollary 2.87 we have that $M_t = M_0 + \int_0^t g_s dW_s$. Finding an explicit expression for the martingale M_t will allow us to identify the process g_s.

a)

$$M_t = \mathbb{E}[Y|\mathcal{F}_t] = \mathbb{E}[W_T|\mathcal{F}_t] = W_t = W_0 + \int_0^t dW_s$$

Then we can identify $g_s = 1$ and write

$$Y = W_T = W_0 + \int_0^T dW_s$$

b)

$$\begin{aligned} M_s &= \mathbb{E}[Y|\mathcal{F}_s] = \int_0^T \mathbb{E}[W_t|\mathcal{F}_s]\,dt \\ &= \int_0^s \mathbb{E}[W_t|\mathcal{F}_s]\,dt + \int_s^T \mathbb{E}[W_t|\mathcal{F}_s]\,dt \\ &= \int_0^s W_t dt + (T-s)\,W_s \end{aligned}$$

By the Paley-Wiener-Zygmund integral (2.4), with $f(s) = s$, we have

$$\int_0^s W_t dt = sW_s - \int_0^s t dW_t$$

Then

$$\begin{aligned} M_s &= sW_s - \int_0^s t dW_t + (T-s)\,W_s = TW_s - \int_0^s t dW_t \\ &= \int_0^s (T-t)\,dW_t \end{aligned}$$

and therefore, $g_t = T - t$, and

$$Y = \int_0^T W_t dt = \int_0^T (T-t)\,dW_t$$

c)

$$\begin{aligned} M_t &= \mathbb{E}[Y|\mathcal{F}_t] = \mathbb{E}\left[W_T^2|\mathcal{F}_t\right] = \mathbb{E}\left[W_T^2 - T|\mathcal{F}_t\right] + \mathbb{E}[T|\mathcal{F}_t] \\ &= W_t^2 - t + T = T + 2\int_0^t W_s dW_s \end{aligned}$$

where we have applied that $W_t^2 - t$ is a martingale and expression (2.16). Therefore, $M_0 = T$, $g_s = 2W_s$, and

$$Y = W_T^2 = T + 2\int_0^T W_s dW_s$$

with $T = \mathbb{E}[Y] = \mathbb{E}\left[W_T^2\right]$.

Chapter 3

1. The process X_t can be expressed as

$$X_t = \int_a^t f(u)\, dW_u = \int_a^s f(u)\, dW_u + \int_s^t f(u)\, dW_u$$
$$= X_s + \int_s^t f(u)\, dW_u$$

By viii) of Proposition 2.13, we know that $\int_s^t f(u)\, dW_u \sim N\left(0, \int_s^t f^2(u)\, du\right)$. Moreover, as f is deterministic, $\int_s^t f(u)\, dW_u$ is independent of $\mathcal{F}_s$, and therefore, the distribution of X_t given $X_s = x$ is $N\left(x, \int_s^t f^2(u)\, du\right)$. Therefore

$$\begin{aligned} P(s, x \,|t, y) &= \mathbb{P}[X_t \leq y \,|X_s = x] \\ &= \frac{1}{\sqrt{2\pi \int_s^t f^2(u)\, du}} \int_{-\infty}^{y} \exp\left\{-\frac{(v-x)^2}{2\int_s^t f^2(u)\, du}\right\} dv \\ &= \frac{1}{\sqrt{2\pi}} \int_{-\infty}^{\frac{y-x}{\sqrt{\int_s^t f^2(u)du}}} e^{-\frac{z^2}{2}}\, dz \\ &= \Phi\left[\frac{y-x}{\sqrt{\int_s^t f^2(u)\, du}}\right] \end{aligned}$$

where Φ is the cumulative distribution function of a standard Gaussian random variable. Now we prove that this transition probability function satisfies the Chapman-Kolmogorov equation:

$$\begin{aligned} &\int_{-\infty}^{+\infty} P(u, z \,|t, y)\, P(s, x \,|u, dz) \\ &= \frac{1}{\sqrt{2\pi \int_s^u f^2(v)\, dv}} \\ &\times \int_{-\infty}^{+\infty} \Phi\left[\frac{y-z}{\sqrt{\int_u^t f^2(v)\, dv}}\right] \phi\left[\frac{z-x}{\sqrt{\int_s^u f^2(v)\, dv}}\right] dz \\ &= \int_{-\infty}^{+\infty} \Phi\left[\frac{y - w\sqrt{\int_s^u f^2(v)\, dv} - x}{\sqrt{\int_u^t f^2(v)\, dv}}\right] \phi[w]\, dw \\ &= \Phi\left[\frac{\frac{y-x}{\sqrt{\int_u^t f^2(v)dv}}}{\sqrt{1 + \frac{\int_s^u f^2(v)dv}{\int_u^t f^2(v)dv}}}\right] \end{aligned}$$

$$= \Phi\left[\frac{y-x}{\sqrt{\int_s^t f^2(v)\,dv}}\right]$$
$$= P(s,x\,|t,y)$$

where we have applied the identity $\int_{-\infty}^{+\infty}\Phi(a+bx)\,\phi(x)\,dx = \Phi\left(\frac{a}{\sqrt{1+b^2}}\right)$ (see Owen (1980), equation 10,010.8), with ϕ the density function of a standard Gaussian random variable.

2. a) Starting from expression (3.18) we have

$$X_t = e^{-\alpha t}X_0 + \mu\left(1-e^{-\alpha t}\right) + \sigma\int_0^t e^{-\alpha(t-u)}\,dW_u$$
$$= \mu + e^{-\alpha t}X_0 - \mu e^{-\alpha t} + \sigma\int_0^s e^{-\alpha(t-u)}\,dW_u$$
$$+\sigma\int_s^t e^{-\alpha(t-u)}\,dW_u$$
$$= \mu + \left[e^{-\alpha s}X_0 - \mu e^{-\alpha s} + \sigma\int_0^s e^{-\alpha(s-u)}\,dW_u\right]e^{-\alpha(t-s)}$$
$$+\sigma\int_s^t e^{-\alpha(t-u)}\,dW_u$$
$$= \mu + (X_s-\mu)\,e^{-\alpha(t-s)} + \sigma\int_s^t e^{-\alpha(t-u)}\,dW_u$$

which coincides with the given expression.

b) Since the integrand in part a) is deterministic, the stochastic integral is independent of $\mathcal{F}_s$. Therefore, by viii) of Proposition 2.13, we have that

$$X_t\,|X_s = x \sim N\left[\mu + (x-\mu)\,e^{-\alpha(t-s)}, \frac{\sigma^2}{2\alpha}\left(1-e^{-2\alpha(t-s)}\right)\right]$$

and then

$$P(s,x\,|t,y) = \mathbb{P}\left[X_t \le y\,|X_s = x\right] = \Phi\left[\frac{y-\mu-(x-\mu)\,e^{-\alpha(t-s)}}{\sigma\sqrt{\frac{1-e^{-2\alpha(t-s)}}{2\alpha}}}\right]$$

3. a) From SDE (3.53) we can write

$$X_t = X_s + \int_s^t \mu ds + \int_s^t \sigma dW_s = X_s + \mu(t-s) + \sigma(W_t - W_s)$$

Thus, we can obtain the transition probability function as follows.

$$
\begin{aligned}
P(s,x\,|t,y) &= \mathbb{P}\left[X_t \leq y\,|X_s = x\right] \\
&= \mathbb{P}\left[W_t - W_s \leq \frac{y - X_s - \mu(t-s)}{\sigma}\,|X_s = x\right] \\
&= \mathbb{P}\left[W_t - W_s \leq \frac{y - x - \mu(t-s)}{\sigma}\right] \\
&= \frac{1}{\sqrt{2\pi(t-s)}} \int_{-\infty}^{\frac{y-x-\mu(t-s)}{\sigma}} e^{-\frac{u^2}{2(t-s)}}\,du \\
&= \frac{1}{\sqrt{2\pi}} \int_{-\infty}^{\frac{y-x-\mu(t-s)}{\sigma\sqrt{t-s}}} e^{-\frac{v^2}{2}}\,dv \\
&= \Phi\left[\frac{y-x-\mu(t-s)}{\sigma\sqrt{t-s}}\right]
\end{aligned}
$$

Taking partial derivative with respect to y, we obtain the transition probability density

$$
p(s,x\,|t,y) = \frac{\partial}{\partial y} P(s,x\,|t,y) = \frac{1}{\sigma\sqrt{t-s}}\phi\left[\frac{y-x-\mu(t-s)}{\sigma\sqrt{t-s}}\right]
$$

b) Using that $\phi'(x) = -x\phi(x)$, we can obtain the partial derivatives of $p(s,x\,|t,y)$, which are given by

$$
\begin{aligned}
\frac{\partial}{\partial t} p(s,x\,|t,y) = &\left\{\frac{-1}{2\sigma(t-s)\sqrt{t-s}} + \frac{\mu\left[y-x-\mu(t-s)\right]}{\sigma^3(t-s)\sqrt{t-s}}\right. \\
&\left. + \frac{\left[y-x-\mu(t-s)\right]^2}{2\sigma^3(t-s)^2\sqrt{t-s}}\right\} \times \phi\left[\frac{y-x-\mu(t-s)}{\sigma\sqrt{t-s}}\right]
\end{aligned}
$$

$$
\frac{\partial}{\partial y} p(s,x\,|t,y) = -\frac{y-x-\mu(t-s)}{\sigma^3(t-s)\sqrt{t-s}} \times \phi\left[\frac{y-x-\mu(t-s)}{\sigma\sqrt{t-s}}\right]
$$

$$
\begin{aligned}
\frac{\partial^2}{\partial y^2} p(s,x\,|t,y) = &\left\{\frac{-1}{\sigma^3(t-s)\sqrt{t-s}} + \frac{\left[y-x-\mu(t-s)\right]^2}{\sigma^5(t-s)^2\sqrt{t-s}}\right\} \\
&\times \phi\left[\frac{y-x-\mu(t-s)}{\sigma\sqrt{t-s}}\right]
\end{aligned}
$$

Replacing these expressions into the left-hand side of the Kolmogorov forward equation (3.43), we obtain

$$\frac{\partial}{\partial t}p(s,x|t,y)+\mu\frac{\partial}{\partial y}p(s,x|t,y)-\frac{\sigma^2}{2}\frac{\partial^2}{\partial y^2}p(s,x|t,y)$$

$$=\left\{\frac{-1}{2\sigma(t-s)\sqrt{t-s}}+\frac{\mu[y-x-\mu(t-s)]}{\sigma^3(t-s)\sqrt{t-s}}+\frac{[y-x-\mu(t-s)]^2}{2\sigma^3(t-s)^2\sqrt{t-s}}\right.$$

$$\left.-\frac{\mu[y-x-\mu(t-s)]}{\sigma^3(t-s)\sqrt{t-s}}+\frac{1}{2\sigma(t-s)\sqrt{t-s}}-\frac{[y-x-\mu(t-s)]^2}{2\sigma^3(t-s)^2\sqrt{t-s}}\right\}$$

$$\times\phi\left[\frac{y-x-\mu(t-s)}{\sigma\sqrt{t-s}}\right]=0$$

On the other hand

$$\begin{aligned}p(s,x|s,y)&=\lim_{t\to s}p(s,x|t,y)=\lim_{t\to s}\frac{1}{\sigma\sqrt{t-s}}\phi\left[\frac{y-x-\mu(t-s)}{\sigma\sqrt{t-s}}\right]\\&=\lim_{t\to s}\frac{1}{\sigma\sqrt{t-s}}\phi\left(\frac{y-x}{\sigma\sqrt{t-s}}\right)\\&=\lim_{t\to s}\frac{1}{\sigma\sqrt{2\pi(t-s)}}e^{-\frac{(y-x)^2}{2\sigma^2(t-s)}}\\&=\delta(y-x)\end{aligned}$$

where we have used that $\lim_{\sigma\to 0}\frac{1}{\sigma\sqrt{2\pi}}e^{-\frac{(x-a)^2}{2\sigma^2}}=\delta(x-a)$.

4. a) Geometric Brownian motion is given by

$$X_t=X_0\exp\left[\left(\mu-\frac{\sigma^2}{2}\right)t+\sigma W_t\right]$$

Taking conditional expectation on X_T and using that $\exp\left(-\frac{\sigma^2}{2}t+\sigma W_t\right)$ is a martingale (see Proposition 1.12), we obtain

$$\begin{aligned}\mathbb{E}[X_T|\mathcal{F}_t]&=X_0e^{\mu T}\mathbb{E}\left[\exp\left(-\frac{\sigma^2}{2}T+\sigma W_T\right)|\mathcal{F}_t\right]\\&=X_0e^{\mu T}\exp\left(-\frac{\sigma^2}{2}t+\sigma W_t\right)\\&=e^{\mu(T-t)}X_t\end{aligned}$$

b) Define $Z_t:=X_0\exp\left(-\frac{\sigma^2}{2}t+\sigma W_t\right)$, so that $X_t=e^{\mu t}Z_t$. The process Z_t is a solution of the GBM SDE with $\mu=0$. Then, we have $dZ_t=\sigma Z_t dW_t$ and

$$Z_t=Z_0+\sigma\int_0^t Z_s dW_s=X_0+\sigma\int_0^t Z_s dW_s$$

Therefore, we can write

$$
\begin{aligned}
X_T &= e^{\mu T} Z_T = e^{\mu T} \left(X_0 + \sigma \int_0^T Z_s dW_s \right) \\
&= e^{\mu T} \left(X_0 + \sigma \int_0^T e^{-\mu s} X_s dW_s \right) \\
&= e^{\mu T} X_0 + \sigma \int_0^T e^{-\mu (T-s)} X_s dW_s
\end{aligned}
$$

which is the representation we were looking for.

c) The transition probability function is

$$P(s,x\,|t,y) = \mathbb{P}[X_t \leq y\,|X_s = x] = \mathbb{P}[\ln X_t \leq \ln y\,|\ln X_s = \ln x]$$

The condition $\ln X_t \leq \ln y$ can be expressed as

$$\ln X_s + \left(\mu - \frac{\sigma^2}{2}\right)(t-s) + \sigma (W_t - W_s) \leq \ln y$$

from where we obtain

$$W_t - W_s \leq \frac{1}{\sigma}\left[\ln\left(\frac{y}{X_s}\right) - \left(\mu - \frac{\sigma^2}{2}\right)(t-s)\right]$$

Then, we have

$$P(s,x\,|t,y) = \frac{1}{\sqrt{2\pi(t-s)}} \int_{-\infty}^{\frac{1}{\sigma}\left[\ln\left(\frac{y}{x}\right) - \left(\mu - \frac{\sigma^2}{2}\right)(t-s)\right]} e^{-\frac{z^2}{2(t-s)}} dz$$

Differentiating with respect to y, we find

$$
\begin{aligned}
p(s,x\,|t,y) &= \frac{\partial}{\partial y} P(s,x\,|t,y) \\
&= \frac{1}{\sigma y \sqrt{2\pi(t-s)}} \exp\left\{ -\frac{\left[\ln\left(\frac{y}{x}\right) - \left(\mu - \frac{\sigma^2}{2}\right)(t-s)\right]^2}{2\sigma^2(t-s)} \right\}
\end{aligned}
$$

d) From (3.20) and Theorem 3.59, we have that GBM is a diffusion with $\rho(t,x) = \mu x$ and $q(t,x) = \sigma^2 x^2$. Then, the Kolmogorov forward PDE is

$$\frac{\partial}{\partial t} p(s,x\,|t,y) + \mu \frac{\partial}{\partial y}[y p(s,x\,|t,y)] - \frac{\sigma^2}{2}\frac{\partial^2}{\partial y^2}\left[y^2 p(s,x\,|t,y)\right] = 0 \tag{4.8}$$

To verify that the transition probability density obtained in c) satisfies the PDE (4.8), we calculate the corresponding partial derivatives.

$$\frac{\partial}{\partial t}p(s,x\,|t,y) = \left[-\frac{1}{2(t-s)} + \frac{\ln^2\left(\frac{y}{x}\right)}{2\sigma^2(t-s)^2} - \frac{\left(\mu-\frac{\sigma^2}{2}\right)}{2\sigma^2}\right] \times p(s,x\,|t,y) \tag{4.9}$$

$$\frac{\partial}{\partial y}[yp(s,x\,|t,y)] = \frac{\left(\mu-\frac{\sigma^2}{2}\right)(t-s)-\ln\left(\frac{y}{x}\right)}{\sigma^2(t-s)}p(s,x\,|t,y)$$

$$\frac{\partial^2}{\partial y^2}\left[y^2p(s,x\,|t,y)\right] = -\frac{1}{\sigma^2(t-s)}\left[\ln\left(\frac{y}{x}\right)-\left(\mu-\frac{\sigma^2}{2}\right)(t-s)+1 - \frac{\left[\ln\left(\frac{y}{x}\right)-\left(\mu-\frac{\sigma^2}{2}\right)(t-s)\right]^2}{\sigma^2(t-s)}\right]p(s,x\,|t,y)$$

From the two last partial derivatives, after some algebra, we obtain

$$\mu\frac{\partial}{\partial y}[yp(s,x\,|t,y)] - \frac{\sigma^2}{2}\frac{\partial^2}{\partial y^2}\left[y^2p(s,x\,|t,y)\right] = -\frac{\ln^2\left(\frac{y}{x}\right)}{2\sigma^2(t-s)^2} + \frac{\left(\mu-\frac{\sigma^2}{2}\right)^2}{2\sigma^2} + \frac{1}{2(t-s)} \times p(s,x\,|t,y)$$

Replacing this result and (4.9) in (4.8) we obtain the desired result.

5. Assume that condition (3.49) holds. Then, if $u(s,x)$ is a solution of the PDE, Theorem 3.48 yields

$$u(s,x) = \mathbb{E}\left[\ln\left(X_T^2\right)|X_s = x\right]$$

where X_t satisfies the geometric Brownian motion SDE

$$dX_t = \mu X_t dt + \sigma X_t dW_t$$

From expression (3.21) we have

$$X_T = X_s \exp\left[\left(\mu-\frac{\sigma^2}{2}\right)(T-s)+\sigma(W_T-W_s)\right]$$

Taking logarithms

$$\ln X_T = \ln X_s + \left(\mu-\frac{\sigma^2}{2}\right)(T-s)+\sigma(W_T-W_s)$$

and then

$$\ln X_T \mid X_s = x \sim N\left(\ln x + \left(\mu - \frac{\sigma^2}{2}\right)(T - s), \sigma^2 (T - s)\right)$$

Finally

$$\begin{aligned} u(s, x) &= \mathbb{E}\left[\ln\left(X_T^2\right) \mid X_s = x\right] = 2\mathbb{E}\left[\ln X_T \mid X_s = x\right] \\ &= 2\left[\ln x + \left(\mu - \frac{\sigma^2}{2}\right)(T - s)\right] \\ &= 2\ln x + \left(2\mu - \sigma^2\right)(T - s) \end{aligned}$$

Appendix A

Measure Theory, Integration, and Functional Analysis

A.1 Measures

Let X be a set and $\mathcal{F}$ a family of subsets of X.

Definition A.1. (σ-algebra) The family $\mathcal{F}$ is a σ-algebra on X if it satisfies the following conditions[1]:

i) $\varnothing \in \mathcal{F}$

ii) If $A \in \mathcal{F}$, then $A^c \in \mathcal{F}$

iii) If $A_n \in \mathcal{F}\ \forall n \in \mathbb{N}$, then $\bigcup_{n=1}^{\infty} A_n \in \mathcal{F}$

iv) If $A_n \in \mathcal{F}\ \forall n \in \mathbb{N}$, then $\bigcap_{n=1}^{\infty} A_n \in \mathcal{F}$

In words, a σ-algebra is a family of subsets of X, which contains the empty set $\varnothing$, and is closed under complements, countable unions, and countable intersections.

Definition A.2. (Measurable space, $\mathcal{F}$-measurable set) The pair $(X, \mathcal{F})$ where X is a set and $\mathcal{F}$ is a σ-algebra on X is called a measurable space. A subset $A \subseteq X$ is called $\mathcal{F}$-measurable if $A \in \mathcal{F}$.

Definition A.3. (Measure) A mapping μ from $\mathcal{F}$ to the extended half-line $[0, \infty]$ is a measure on a measurable space $(X, \mathcal{F})$ if it satisfies:

i) $\mu(\varnothing) = 0$

ii) (Countable additivity) If $A_n \in \mathcal{F}\ \forall n \in \mathbb{N}$ and $A_i \cap A_j = \varnothing$ for $i \neq j$, then

$$\mu\left(\bigcup_{n=1}^{\infty} A_n\right) = \sum_{n=1}^{\infty} \mu(A_n)$$

Definition A.4. (Measure space) A measure space is a triple $(X, \mathcal{F}, \mu)$, where μ is a measure on the measurable space $(X, \mathcal{F})$.

[1] Actually, iv) follows from ii) and iii). See Axler (2020), 2.25.

A.2 Integration

Given a measure space $(X, \mathcal{F}, \mu)$, our aim now is to define the integral of a function $f : X \longrightarrow \mathbb{R}$ with respect to the measure $\mu : \mathcal{F} \longrightarrow [0, \infty]$. In order to do that, we will follow several steps. The first one is to define the integral of simple functions.

Definition A.5. (Simple function) A function $f : X \longrightarrow \mathbb{R}$ is called simple if it can be written as

$$f(x) = \sum_{n=1}^{k} c_n \mathbf{1}_{A_n}(x) \tag{A.1}$$

for every $x \in X$, where $c_1, \ldots, c_k \in \mathbb{R}$, $A_1, \ldots, A_k$ are $\mathcal{F}$-measurable sets, and $\mathbf{1}_A$ is the indicator function of the set A, defined by

$$\mathbf{1}_A(x) = \begin{cases} 1 & \text{if} \quad x \in A \\ 0 & \text{if} \quad x \in A^c \end{cases}$$

Definition A.6. (Integral of a simple function) Consider a simple function f of the form (A.1). The integral of f is defined as

$$\int_X f(x)\, d\mu(x) = \sum_{n=1}^{k} c_n \mu(A_n)$$

With this first definition of integral we can now calculate some basic integrals.

Proposition A.7. *Let $(X, \mathcal{F}, \mu)$ be a measure space. Then*

i) $\int_X 0 d\mu(x) = 0$

ii) $\int_X k d\mu(x) = k\mu(X), \;\; k \in \mathbb{R} - \{0\}$

Proof. For the first integral, it suffices to write $0 = \mathbf{1}_\varnothing(x)$ and to apply Definition A.6 and i) of Definition A.3. For the second one, it is enough to consider that $k = k\mathbf{1}_X(x)$ and to apply Definition A.6. □

Note that in ii) of the previous proposition, the result is finite if, and only if, $\mu(X) < \infty$.

In order to extend the definition of integral to more general functions, we need to consider measurable functions.

Definition A.8. (Measurable function) A function $f : X \longrightarrow \mathbb{R}$ is $\mathcal{F}$-measurable or simply measurable, if $f^{-1}((a, \infty)) \in \mathcal{F} \;\; \forall a \in \mathbb{R}$, i.e., if the set

$$\{x \in X : f(x) > a\}$$

is measurable.

As a simple consequence of the previous definition, any constant function $f : X \longrightarrow \mathbb{R}$ such that $f(x) = k \;\; \forall x \in X$, is measurable. To prove this, it is enough to consider that if $k \leq a$, then $f^{-1}((a, \infty)) = \varnothing$ and if $k > a$, then $f^{-1}((a, \infty)) = X$, and $\varnothing, X \in \mathcal{F}$.

There are other equivalent conditions in order to determine if a function is measurable.

Proposition A.9. *The following statements are equivalent:*

i) f *is* $\mathcal{F}$*-measurable.*

ii) $\{x \in X : f(x) \leq a\} \in \mathcal{F} \quad \forall a \in \mathbb{R}$

iii) $\{x \in X : f(x) > a\} \in \mathcal{F} \quad \forall a \in \mathbb{R}$

iv) $\{x \in X : f(x) \geq a\} \in \mathcal{F} \quad \forall a \in \mathbb{R}$

Proof. See Royden and Fitzpatrick (2010), Proposition 1, page 54. □

The property of being measurable goes well with elementary operations between functions.

Proposition A.10. *Let* f, g *be two* $\mathcal{F}$*-measurable functions. Then*

i) $f + g$, $f - g$, *and* fg *are* $\mathcal{F}$*-measurable functions.*

ii) If $g(x) \neq 0 \ \forall x \in X$, *then* $\dfrac{f}{g}$ *is* $\mathcal{F}$*-measurable.*

Proof. See Axler (2020), 2.46. □

Note that by i) of the previous proposition, we have that $\alpha f + \beta g$ is also a measurable function for every $\alpha, \beta \in \mathbb{R}$.

The pointwise limit of a sequence of measurable functions is also a measurable function.

Proposition A.11. *Let* $\{f_n\}_{n=1}^{\infty}$ *be a sequence of* $\mathcal{F}$*-measurable functions such that* $\lim_{n\to\infty} f_n(x)$ *exists for each* $x \in X$. *Then, the function* f *defined for each* $x \in X$ *by*

$$f(x) = \lim_{n\to\infty} f_n(x)$$

is an $\mathcal{F}$*-measurable function.*

Proof. See Axler (2020), 2.48. □

Our next step towards a general definition of the integral is to define it for non-negative functions.

Definition A.12. (Integral of a non-negative function) Let $f : X \to \mathbb{R}$ be $\mathcal{F}$-measurable and non-negative. Then, the integral of f is defined as

$$\int_X f(x)\, d\mu(x) = \sup_{\varphi} \int_X \varphi(x)\, d\mu(x)$$

where the supremum is taken over those simple functions φ such that $0 \leq \varphi \leq f$.

The following result allows us to obtain the measure of a measurable set by integration.

Proposition A.13. *If* $A \in \mathcal{F}$, *then*

$$\mu(A) = \int_X \mathbf{1}_A(x)\, d\mu(x)$$

Proof. See Axler (2020), 3.4. □

Definition A.14. (Positive and negative part of a function) Given a function $f : X \to \mathbb{R}$, its positive part, f^+, is the function defined by $f^+ = \max(f, 0)$, and its negative part, f^-, is the function defined by $f^- = \max(-f, 0)$. Note that f^+ and f^- are both non-negative functions satisfying $f = f^+ - f^-$ and $|f| = f^+ + f^-$

Proposition A.15. *If $f : X \to \mathbb{R}$ is $\mathcal{F}$-measurable, then f^+, f^-, and $|f|$ are $\mathcal{F}$-measurable.*

Proof. See Halmos (1950), page 82. □

We are now in a position to present the general definition of the integral of a function.

Definition A.16. (Integral of a function) If $f : X \to \mathbb{R}$ is an $\mathcal{F}$-measurable function, the integral of f is defined as

$$\int_X f(x)\, d\mu(x) = \int_X f^+(x)\, d\mu(x) - \int_X f^-(x)\, d\mu(x)$$

Definition A.17. (Integrable function) Let $f : X \to \mathbb{R}$ be an $\mathcal{F}$-measurable function. We say that f is integrable if

$$\int_X |f(x)|\, d\mu(x) < \infty$$

If f is integrable we will write $f \in L^1(X, \mathcal{F}, \mu)$ or simply $f \in L^1$.

From the two previous definitions, we have the following possibilities for the values of the integral:

- If $\int_X f^+(x)\, d\mu(x) < \infty$ and $\int_X f^-(x)\, d\mu(x) < \infty$, then f is integrable and its integral is finite.
- If $\int_X f^+(x)\, d\mu(x) = \int_X f^-(x)\, d\mu(x) = \infty$, then f is not integrable, and the integral cannot be defined ($\infty - \infty$).
- If $\int_X f^+(x)\, d\mu(x) = \infty$ and $\int_X f^-(x)\, d\mu(x) < \infty$, then f is not integrable and $\int_X f(x)\, d\mu(x) = \infty$.
- If $\int_X f^+(x)\, d\mu(x) < \infty$ and $\int_X f^-(x)\, d\mu(x) = \infty$, then f is not integrable and $\int_X f(x)\, d\mu(x) = -\infty$.

Definition A.18. (Integrable function over a measurable set, integral of a function over a measurable set) Let $f : X \to \mathbb{R}$ be an $\mathcal{F}$-measurable function and $A \subset X$ a measurable set. We say that f is integrable over A if

$$\int_X |f(x)\, \mathbf{1}_A(x)|\, d\mu(x) < \infty$$

If f is integrable over A, we define the integral of f over A as

$$\int_A f(x)\, d\mu(x) = \int_X f(x)\, \mathbf{1}_A(x)\, d\mu(x)$$

The following proposition contains the first properties of the integral of a function.

Proposition A.19. *If* $f, g \in L^1(X, \mathcal{F}, \mu)$, *the following relations hold:*

i) $\forall \alpha, \beta \in \mathbb{R}$

$$\int_X [\alpha f(x) + \beta g(x)]\, d\mu(x) = \alpha \int_X f(x)\, d\mu(x) + \beta \int_X g(x)\, d\mu(x)$$

ii) If $f(x) \leq g(x) \quad \forall x \in X$, *then*

$$\int_X f(x)\, d\mu(x) \leq \int_X g(x)\, d\mu(x)$$

iii) $\left|\int_X f(x)\, d\mu(x)\right| \leq \int_X |f(x)|\, d\mu(x)$

iv) If $A, B \in \mathcal{F}$ *with* $A \cap B = \varnothing$, *then*

$$\int_{A \cup B} f(x)\, d\mu(x) = \int_A f(x)\, d\mu(x) + \int_B f(x)\, d\mu(x)$$

Proof. See Axler (2020) 3.20-3.23 and Royden and Fitzpatrick (2010), Theorem 12, page 373. □

The following theorems state under which conditions we can interchange limits and integrals.

Theorem A.20. ***(Monotone convergence)*** *Let* $\{f_n\}_{n=1}^{\infty}$ *be a non-negative and increasing sequence of measurable functions defined on* X, *such that* $\lim_{n\to\infty} f_n(x)$ *exists for each* $x \in X$. *Then, if we define the function* $f : X \to \mathbb{R}$ *by*

$$f(x) = \lim_{n\to\infty} f_n(x)$$

we have that

$$\int_X f(x)\, d\mu(x) = \lim_{n\to\infty} \int_X f_n(x)\, d\mu(x)$$

Proof. See Axler (2020), 3.11. □

Theorem A.21. ***(Dominated convergence)*** *Let* $\{f_n\}_{n=1}^{\infty}$ *be a sequence of measurable functions such that*

$$\lim_{n\to\infty} f_n(x) = f(x) \quad \forall x \in X$$

for some function f. *If there exists a non-negative function* $g \in L^1$ *satisfying*

$$|f_n(x)| \leq g(x) \quad \forall n \in \mathbb{N} \quad \mu - a.e. \text{ on } X$$

then f *is integrable over* X *and*

$$\int_X f(x)\, d\mu(x) = \lim_{n\to\infty} \int_X f_n(x)\, d\mu(x)$$

Proof. See Royden and Fitzpatrick (2010), page 376. □

The following definition gives a name to those sets whose measure is zero.

Definition A.22. (Null set) A measurable set N in the measure space $(X, \mathcal{F}, \mu)$ is a null set if, and only if, $\mu(N) = 0$.

Definition A.23. (Almost everywhere) If a certain property holds for every $x \in X$ except on a null set, then we say that the property holds almost everywhere, and it is often abbreviated as a.e. or μ-a.e.

Theorem A.24. *Let $f, g : X \to \mathbb{R}$ be two integrable functions. Then the following properties hold:*

i) If N is a null set, then

$$\int_N f(x)\, d\mu(x) = 0$$

ii) $\int_A f(x)\, d\mu(x) = 0 \ \ \forall A \in \mathcal{F} \Rightarrow f = 0 \quad \mu\text{-a.e.}$

iii) $\int_A f(x)\, d\mu(x) = \int_A g(x)\, d\mu(x) \ \ \forall A \in \mathcal{F} \Leftrightarrow f = g \quad \mu\text{-a.e.}$

Proof. For i) see Halmos (1950), Theorem C, page 104, for ii) see Rudin (1987), Theorem 1.39 b), iii) is a direct consequence of i) and ii). □

The following concepts and results will be useful in a probabilistic framework. Let $(X, \mathcal{F}, \mu)$ be a measure space, and $f : X \to \mathbb{R}$ a measurable function. Define the mapping $\mu \circ f^{-1} : \mathcal{B}(\mathbb{R}) \to [0, \infty]$ by

$$\left(\mu \circ f^{-1}\right)(A) = \mu\left(f^{-1}(A)\right) \quad \forall A \in \mathcal{B}(\mathbb{R})$$

Proposition A.25. *The mapping $\mu \circ f^{-1}$ is a measure on $(\mathbb{R}, \mathcal{B}(\mathbb{R}))$.*

Proof. See Cohn (2013), pages 75-76. □

Definition A.26. (Image of a measure under a function) The measure $\mu \circ f^{-1}$ is called the image of μ under f.

Proposition A.27. *Let $(X, \mathcal{F}, \mu)$ be a measure space, and $f : X \to \mathbb{R}$ a measurable function. Let $g : \mathbb{R} \to \mathbb{R}$ be a Borel measurable function. Then*

$$g \circ f \in L^1(X, \mathcal{F}, \mu) \Leftrightarrow g \in L^1\left(\mathbb{R}, \mathcal{B}(\mathbb{R}), \mu \circ f^{-1}\right)$$

Moreover, if these functions are integrable, then

$$\int_X (g \circ f)(x)\, d\mu(x) = \int_{\mathbb{R}} g(x)\, d\left(\mu \circ f^{-1}\right)(x)$$

Proof. See Cohn (2013), Proposition 2.6.8. □

A.3 σ-Algebras and Generators

In this section we try to answer the following question. Given a set X and a family $\mathcal{S}$ of subsets of X, what is the "smallest" σ-algebra containinng $\mathcal{S}$? The response appears in the following result.

Proposition A.28. *Let $\mathcal{S}$ be any family of subsets of X, and let $\mathcal{G}$ be the intersection of all the σ-algebras on X that contain $\mathcal{S}$. Then, the set $\mathcal{G}$ satisfies the following properties:*

i) *$\mathcal{G}$ extends $\mathcal{S}$, i.e., $\mathcal{S} \subset \mathcal{G}$*

ii) *$\mathcal{G}$ is a σ-algebra on X.*

iii) *$\mathcal{G}$ is the smallest σ-algebra containing $\mathcal{S}$, in the sense that if $\mathcal{K}$ is any σ-algebra containing $\mathcal{S}$, then $\mathcal{G} \subseteq \mathcal{K}$.*

Proof. See Cohn (2013), Corollary 1.1.3. □

Definition A.29. (σ-algebra generated by a family of subsets, generator system) The σ-algebra $\mathcal{G}$ of Proposition A.28 is called the σ-algebra generated by $\mathcal{S}$, and is denoted by $\mathcal{G} = \sigma(\mathcal{S})$. The class $\mathcal{S}$ is called a generator system for $\mathcal{G}$.

A.4 Product Measures

Let $(X, \mathcal{F}, \mu)$ and $(Y, \mathcal{G}, \nu)$ be two measure spaces. Our aim in this section is to define a measure on the Cartesian product $X \times Y$ and to find out how we can calculate integrals of functions on $X \times Y$ with respect to that measure.

Definition A.30. (Measurable rectangle) A subset Z of $X \times Y$ is a measurable rectangle if it is of the form $Z = A \times B$ where $A \in \mathcal{F}$ and $B \in \mathcal{G}$.

Definition A.31. (Product σ-algebra) The σ-algebra generated by all measurable rectangles on $X \times Y$ is called the product σ-algebra on $X \times Y$, and is denoted by $\mathcal{F} \otimes \mathcal{G}$.

In order to state the main result on product measures, we need to work with a special type of measure spaces, known as σ-finite measure spaces.

Definition A.32. (σ-finite measure, σ-finite measure space) Let $(X, \mathcal{F}, \mu)$ be a measure space. We say that μ is a σ-finite measure if X is the union of a sequence of measurable sets with finite measure. If μ is σ-finite, $(X, \mathcal{F}, \mu)$ is a σ-finite measure space.

Theorem A.33. *Let $(X, \mathcal{F}, \mu)$ and $(Y, \mathcal{G}, \nu)$ be two σ-finite measure spaces. Then, there exists a unique measure $\mu \times \nu$ on $(X \times Y, \mathcal{F} \otimes \mathcal{G})$ such that*

$$(\mu \times \nu)(A \times B) = \mu(A)\,\nu(B)$$

for every measurable rectangle $A \times B$.

Proof. See Cohn (2013), Theorem 5.14. □

Definition A.34. (Product measure) The measure $\mu \times \nu$ introduced in Theorem A.33 is called the product measure of measures μ and ν.

The following theorem shows that, under some conditions, we can obtain an integral with respect to the product measure $\mu \times \nu$ by calculating iterated integrals with respect to μ and ν in any order.

Theorem A.35. ***(Fubini's Theorem)*** *Let $(X, \mathcal{F}, \mu)$ and $(Y, \mathcal{G}, \nu)$ be two σ-finite measure spaces, and let $f : X \times Y \to \mathbb{R}$ be an $\mathcal{F} \otimes \mathcal{G}$-measurable and $\mu \times \nu$ integrable function. Then*

$$\int_{X\times Y} f(x,y)\, d(\mu\times\nu)(x,y) = \int_X \left[\int_Y f(x,y)\, d\nu(y)\right] d\mu(x)$$
$$= \int_Y \left[\int_X f(x,y)\, d\mu(x)\right] d\nu(y)$$

Proof. See Cohn (2013), Theorem 5.2.2. □

A.5 The Borel σ-algebra

In this section we illustrate the previous concepts with the example of the most important σ-algebra, whose definition is as follows.

Definition A.36. (Borel σ-algebra, Borel sets) The Borel σ-algebra on $\mathbb{R}$, $\mathcal{B}(\mathbb{R})$, is the σ-algebra generated by the class of open sets of $\mathbb{R}$. The elements of this σ-algebra are called Borel sets. An analogous definition holds for the Borel σ-algebra on $\mathbb{R}^n$, $\mathcal{B}(\mathbb{R}^n)$.[2]

The Borel σ-algebra is a surprisingly intricate object as can be seen from the following two properties:

- There is no finite procedure involving countable unions, countable intersections, and complements, for constructing the Borel σ-algebra.[3]
- There are subsets of $\mathbb{R}$ that are not Borel sets, but are highly non-trivial (see Royden and Fitzpatrick (2010), Proposition 22, page 53).

This last property means that for most subsets of $\mathbb{R}$ arising in applications, it is unnecessary to check explicitly whether they are Borel sets. All subsets of $\mathbb{R}$ that appear in usual applications are Borel sets, as the following examples demonstrate (Axler (2020), 2.30).

- Every open subset of $\mathbb{R}$ is a Borel set by the definition of the Borel σ-algebra.
- Every closed subset of $\mathbb{R}$ is a Borel set, because A^c is an open subset.
- Every countable subset $B = \{x_1, x_2, \ldots\} \subset \mathbb{R}$ is a Borel set, since $B = \bigcup_{n=1}^{\infty} \{x_n\}$, and each $\{x_n\}$ is a closed set.
- Every half-open interval $[a, b)$ is a Borel set because $[a, b) = \bigcap_{n=1}^{\infty} \left(a - \frac{1}{n}, b\right)$.

[2]Recall that a set $A \subseteq \mathbb{R}$ is open if for every $x \in A$, there exists an $\varepsilon > 0$ such that $(x - \varepsilon, x + \varepsilon) \subset A$. $B \subseteq \mathbb{R}$ is closed if B^c is open. For the definitions of open and closed sets in more general spaces, see Section A.8.

[3]This property is a consequence of a general result in the study of Descriptive Set Theory, which is known as the Borel Hierarchy Non-Collapsibility Theorem (see Kechris (1995), Theorem 22.4).

A.6 The Lebesgue Measure

Now that we have a σ-algebra defined on $\mathbb{R}$, our aim is to find a measure on $\mathbb{R}$ that extends the concept of length of an interval. The best candidate is the outer measure.

Definition A.37. (Outer measure) The outer measure $|A|$ of a set $A \subset \mathbb{R}$ is defined by

$$|A| = \inf \left\{ \sum_{n=1}^{\infty} l(I_n) : I_n \text{ are open intervals such that } A \subset \bigcup_{n=1}^{\infty} I_n \right\}$$

where $l(I_n)$ is the length of the interval I_n.

The outer measure has some of the properties that we are looking for.

Proposition A.38. *The outer measure satisfies the following properties:*

i) Every countable set has outer measure 0.

ii) If $A, B \subset \mathbb{R}$, with $A \subset B$, then $|A| \leq |B|$.

iii) If $t \in \mathbb{R}$ and $A \subset \mathbb{R}$, then $|t + A| = |A|$.

iv) If $a, b \in \mathbb{R}$, with $a < b$, then $|[a, b]| = |(a, b)| = b - a$.

Proof. See Axler (2020), 2.4, 2.5, 2.7, 2.14, and exercise 6 of Chapter 2. □

Despite its name, the outer measure is not a measure on the σ-algebra of all subsets of $\mathbb{R}$, because it fails to satisfy countable additivity.

Proposition A.39. *There exist $A, B \subset \mathbb{R}$ with $A \cap B = \emptyset$ such that $|A \cup B| \neq |A| + |B|$.*

Proof. See Axler (2020), 2.18. □

This result does not mean that outer measure is a bad candidate for a measure. In fact, it can be proven that there is no mapping from the set of all subsets of $\mathbb{R}$ to $[0, \infty]$ that satisfies iii) and iv) of Proposition A.38, and countable additivity (see Axler (2020), 2.22).

Therefore, the only solution is to reduce the "size" of the σ-algebra so that the outer measure satisfies countable additivity. As the following result states, this is achieved by considering the Borel σ-algebra.

Theorem A.40. *Outer measure is a measure on the measurable space* $(\mathbb{R}, \mathcal{B}(\mathbb{R}))$.

Proof. See Axler (2020), 2.68. □

Now we can define the measure we are looking for.

Definition A.41. (Lebesgue measure, Lebesgue integral) The Lebesgue measure λ is the outer measure restricted to the Borel σ-algebra, $\lambda : \mathcal{B}(\mathbb{R}) \to [0, \infty]$, $\lambda(A) = |A|$. The integral with respect to the Lebesgue measure is known as Lebesgue integral.

All the general results we obtained previously about the integral of a measurable function are directly applicable to the Lebesgue integral.

The following result shows that the Lebesgue integral generalizes the Riemann integral.

Theorem A.42. *If f is Riemann integrable on $[a,b]$, then f is Lebesgue integrable on $[a,b]$ and both integrals coincide:*

$$\int_a^b f(x)\,dx = \int_{[a,b]} f(x)\,d\lambda(x)$$

Proof. See Axler (2020), 3.34. □

Sometimes, by an abuse of notation, we will express the Lebesgue integral of the function f on $[a,b]$ as $\int_a^b f(x)\,dx$.

A.7 L^p Spaces

Our goal now is to provide the set of measurable functions with a tool that allows us to somehow measure the "distance" between two functions, similar to how distances are measured in Euclidean space. To do this, we need the following definition.

Definition A.43. (Norm, normed vector space) A norm on a vector space V is a mapping $\|\cdot\| : V \longrightarrow [0,\infty)$ such that:

i) $\|f\| = 0 \Leftrightarrow f = 0$

ii) $\|\alpha f\| = |\alpha|\,\|f\| \quad \forall \alpha \in \mathbb{R} \;\; \forall f \in V$

iii) $\|f+g\| \le \|f\| + \|g\| \quad \forall f,g \in V$

A normed vector space is a pair $(V, \|\cdot\|)$, where V is a vector space and $\|\cdot\|$ is a norm on V.

Given a measure space $(X, \mathcal{F}, \mu)$, the set of measurable functions from X to $\mathbb{R}$ is a vector space, as a subspace of the vector space of functions from X to $\mathbb{R}$ (see i) of Proposition A.10 and subsequent comment). Our candidate for a norm in this space is the following.

Definition A.44. (p-norm, essential supremum) Let $f : X \longrightarrow \mathbb{R}$ be an $\mathcal{F}$-measurable function, and $p \in (0,\infty)$. The p-norm of f, $\|f\|_p$, is defined by

$$\|f\|_p = \left[\int_X |f(x)|^p\,d\mu(x)\right]^{1/p} \tag{A.2}$$

and the essential supremum of f, $\|f\|_\infty$, is defined by

$$\|f\|_\infty = \inf\{t > 0 : \mu(\{x \in X : |f(x)| > t\}) = 0\}$$

Despite its name, $\|\cdot\|_p$, $0 < p \le \infty$, is not a norm in the set of $\mathcal{F}$-measurable functions, because if $E \in \mathcal{F}$, $E \neq \varnothing$, with $\mu(E) = 0$, then $\|\mathbf{1}_E\|_p = 0$, but $\mathbf{1}_E \neq 0$.

In order to make $\|\cdot\|_p$ a norm, we have to transform the space itself. Consider the binary relation on the set of $\mathcal{F}$-measurable functions determined by

$$f \sim g \Leftrightarrow f(x) = g(x) \quad \mu - a.e. \Leftrightarrow \mu(\{x \in X : f(x) \neq g(x)\}) = 0$$

This relation satisfies the properties of reflexivity, symmetry and transitivity, and then it is an equivalence relation. Therefore, it induces a partition in the set of $\mathcal{F}$-measurable functions, resulting in a disjoint collection of equivalence classes. We denote by $[f]$ the equivalence class that contains the function f.

Definition A.45. **($L^p(\mu)$ space)** Let $p \in (0, \infty]$. We define $L^p(\mu)$ as the collection of equivalence classes $[f]$, such that $\|f\|_p < \infty$, where f is any representative of the class $[f]$.

Note that the definition of $L^p(\mu)$ is independent of the representative of the class, since different representatives differ only on a subset of measure zero, and then, they have the same p-norm.

Proposition A.46. *The set $L^p(\mu)$ with the addition and scalar multiplication defined, for $[f], [g] \in L^p(\mu)$, and $\alpha \in \mathbb{R}$, by*

$$[f] + [g] = [f + g]$$
$$\alpha [f] = [\alpha f]$$

is a vector space for $1 \le p \le \infty$.

Proof. See, Royden and Fitzpatrick (2010), page 136. □

Now that we have the appropriate vector space, $L^p(\mu)$, we need to define how the p-norm acts on its elements.

Definition A.47. **(p-norm on $L^p(\mu)$)** If $[f] \in L^p(\mu)$, then its p-norm is given by

$$\|[f]\|_p = \|f\|_p$$

where f is any representative of the class $[f]$.

From now on, as usual, we denote the equivalence classes of the relation $\sim$ by f rather than $[f]$, but having in mind that $f = g$ means that $f(x)$ differs from $g(x)$ on a set of measure zero.

To prove that $\|\cdot\|_p$ is a norm on $L^p(\mu)$, it is necessary to use some inequalities that are of great interest in themselves. We present them in the results below.

Theorem A.48. ***(Hölder's inequality)*** *Let $p \in [1, \infty)$, and q, determined by $\frac{1}{p} + \frac{1}{q} = 1$, be the conjugate exponent of p. If $f \in L^p(\mu)$ and $g \in L^q(\mu)$, then fg is integrable and satisfies*

$$\|fg\|_1 \le \|f\|_p \|g\|_q$$

Proof. See, Royden and Fitzpatrick (2010), Theorem 1, page 140. □

Theorem A.49. ***(Minkowski's inequality)*** *Let $p \in [1, \infty]$. If $f, g \in L^p(\mu)$, then*

$$\|f + g\|_p \le \|f\|_p + \|g\|_p$$

Proof. See, Royden and Fitzpatrick (2010), page 141. □

Proposition A.50. *$\|\cdot\|_p$ is a norm on $L^p(\mu)$ for $1 \le p \le \infty$. Therefore, $\left(L^p(\mu), \|\cdot\|_p\right)$ is a normed vector space.*

Proof. The result follows directly from Definitions A.43, A.44, A.45, and Minkowski's inequality. □

The following result states the inclusion relation between $L^p(\mu)$ spaces.

Proposition A.51. *If $(X, \mathcal{F}, \mu)$ is a measure space with $\mu(X) < \infty$ and $0 < p < q < \infty$, then $L^q(\mu) \subset L^p(\mu)$.*

Proof. See, Axler (2020), 7.10. □

L^p spaces possess an important mathematical property, which is that of being Banach spaces. Before presenting this result in formal terms, we need some preliminary definitions.

Definition A.52. (Convergent sequence) A sequence $\{f_n\}_{n=1}^{\infty}$ in a normed vector space $(V, \|\cdot\|)$ is said to converge to $f \in V$, provided that

$$\lim_{n\to\infty} \|f_n - f\| = 0$$

In this case, we write $\lim_{n\to\infty} f_n = f$.

Definition A.53. (Cauchy sequence) A sequence $\{f_n\}_{n=1}^{\infty}$ in a normed vector space $(V, \|\cdot\|)$ is a Cauchy sequence if for each $\varepsilon > 0$, there exists $n_\varepsilon \in \mathbb{N}$ such that $\|f_n - f_m\| < \varepsilon$ for all $m, n \geq n_\varepsilon$.

Proposition A.54. *Every convergent sequence in a normed vector space is a Cauchy sequence.*

Proof. See, Royden and Fitzpatrick (2010), Proposition 4, page 145. □

As the previous result states, for sequences in a normed vector space, the convergence property implies the Cauchy property. The converse result does not hold in general. Spaces in which it is satisfied have a special name.

Definition A.55. (Complete space, Banach space) A normed vector space $(V, \|\cdot\|)$ is said to be complete if every Cauchy sequence in V converges to an element of V. A Banach space is a complete normed vector space.

Theorem A.56. *$L^p(\mu)$ is a Banach space for $1 \leq p \leq \infty$.*

Proof. See, Axler (2020), 7.24. □

A.8 Hilbert Spaces

Hilbert spaces are a class of Banach spaces with special properties that make them useful in many applications.

The following definition is a generalization of the usual scalar product of Euclidean space.

Definition A.57. (Inner product, inner product space) An inner product on a vector space V is a mapping $\langle\cdot,\cdot\rangle : V \times V \to \mathbb{R}$ that satisfies:

i) $\langle f, f\rangle \geq 0 \ \forall f \in V$

ii) $\langle f, f\rangle = 0 \Leftrightarrow f = 0$

iii) $\langle \alpha f + \beta g, h\rangle = \alpha\langle f, h\rangle + \beta\langle g, h\rangle \quad \forall f, g, h \in V \quad \forall \alpha, \beta \in \mathbb{R}$

iv) $\langle f, g\rangle = \langle g, f\rangle \quad \forall f, g \in V$

A vector space with an inner product defined on it is called an inner product space.

The following result contains some basic properties of the inner product.

Proposition A.58. *If $(V, \langle\cdot,\cdot\rangle)$ is an inner product space, then*

i) $\langle 0, f\rangle = \langle f, 0\rangle = 0 \;\; \forall f \in V$

ii) $\langle f, \alpha g + \beta h\rangle = \alpha \langle f, g\rangle + \beta \langle f, h\rangle \;\; \forall f, g, h \in V \;\; \forall \alpha, \beta \in \mathbb{R}$

Proof. See Axler (2020), 8.3. □

Starting with an inner product, it is possible to define a norm, which transforms each inner product space into a normed vector space.

Proposition A.59. *Let $(V, \langle\cdot,\cdot\rangle)$ be an inner product space. Then, the mapping $||\cdot|| : V \to \mathbb{R}$, defined for each $f \in V$ by*

$$||f|| = \sqrt{\langle f, f\rangle} \tag{A.3}$$

is a norm on V and the pair $(V, ||\cdot||)$ is a normed vector space.

Proof. See, Axler (2020), 8.19. □

The inner product can be recovered from the norm induced by it, according to the following result.

Proposition A.60. ***(Polarization identity)*** *If $(V, \langle\cdot,\cdot\rangle)$ is an inner product space and $||\cdot||$ is the norm induced by $\langle\cdot,\cdot\rangle$, then*

$$\langle f, g\rangle = \frac{1}{4}\left[||f+g||^2 - ||f-g||^2\right] \;\; \forall f, g \in V \tag{A.4}$$

The following theorem states a necessary and sufficient condition for a norm to originate from an inner product.

Theorem A.61. ***(Jordan-von Neumann)*** *A normed vector space $(V, ||\cdot||)$ is an inner product space if, and only if, the norm $||\cdot||$ satisfies the parallelogram identity*

$$||f+g||^2 + ||f-g||^2 = 2\left(||f||^2 + ||g||^2\right) \tag{A.5}$$

Proof. See the original paper, Jordan and Neumann (1935), Theorem 1. □

The following definition is a generalization of the concept of perpendicularity to inner product spaces.

Definition A.62. (Orthogonal vectors) Let $(V, \langle\cdot,\cdot\rangle)$ be an inner product space. Two vectors $f, g \in V$ are said to be orthogonal if $\langle f, g\rangle = 0$.

Theorem A.63. ***(Pythagorean Theorem)*** *Let f, g be two orthogonal vectors of an inner product space, and let $||\cdot||$ be the norm induced by the inner product. Then*

$$||f+g||^2 = ||f||^2 + ||g||^2$$

Proof. See Axler (2020), 8.9. □

Theorem A.64. ***(Cauchy-Schwarz inequality)*** *If f, g are vectors of an inner product space, then*

$$|\langle f, g\rangle| \leq \|f\| \, \|g\|$$

The equality holds if, and only if, one of f, g is a scalar multiple of the other.

Proof. See Axler (2020), 8.11. □

Definition A.65. (Orthogonal complement) Let $(V, \langle\cdot,\cdot\rangle)$ be an inner product space, and $U \subseteq V$. The orthogonal complement of U, denoted by $U^{\perp}$, is the set determined by

$$U^{\perp} = \{h \in V : \langle g, h\rangle = 0 \;\; \forall g \in U\}$$

i.e., $U^{\perp}$ contains all the vectors of V that are orthogonal to every vector of U.

In order to present the properties of the orthogonal complement, we need to add a metric structure to the space V.

Definition A.66. (Distance, metric space) A distance on a non-empty set S is a mapping $d : S \times S \to [0, \infty)$ that satisfies:

i) $\mathrm{d}(x, x) = 0 \;\; \forall x \in S$

ii) If $x, y \in S$ and $d(x, y) = 0$, then $x = y$.

iii) $\mathrm{d}(x, y) = d(y, x) \;\; \forall x, y \in S$

iv) $\mathrm{d}(x, z) \leq d(x, y) + d(y, z) \;\; \forall x, y, z \in S$

The pair (S, d), where S is a set and d a distance on S is called a metric space.

The notion of distance allows us to define limits of sequences in metric spaces.

Definition A.67. (Convergent sequence, limit of a sequence) Let (S, d) be a metric space, $x \in S$, and $\{x_n\}_{n=1}^{\infty}$ a sequence of elements of S. We say that the sequence $\{x_n\}_{n=1}^{\infty}$ converges to x, denoted by $\lim_{n\to\infty} x_n = x$, if $\lim_{n\to\infty} d(x_n, x) = 0$. In this case, we say that x is the limit of the sequence $\{x_n\}_{n=1}^{\infty}$.

Now we introduce some definitions related to the topology of metric spaces.

Definition A.68. (Open ball) Let (S, d) be a metric space and let x be an element of S. The open ball centered at x with radius $r \in \mathbb{R}^{+}$, denoted by $B(x, r)$ is the set defined by

$$B(x, r) = \{y \in S : d(x, y) < r\}$$

Definition A.69. (Open set, closed set) Let (S, d) be a metric space, and $M, N \subseteq S$. M is called open if for every $x \in M$, there exists $r > 0$ such that $B(x, r) \subset M$. N is called closed if its complement in S is open.

Definition A.70. (Closure) Let (S, d) be a metric space and $E \subseteq S$. The closure of E, denoted by $\overline{E}$ is the set defined by

$$\overline{E} = \{y \in S : B(y, \varepsilon) \cap E \neq \varnothing \;\; \forall \varepsilon > 0\}$$

Definition A.71. (Dense set) Let (S, d) be a metric space, and $E \subseteq S$. We say that E is dense in S if $\overline{E} = S$.

The following proposition contains some important properties of the closure of a set.

Proposition A.72. *Let (S, d) be a metric space and $E \subseteq S$. Then, the following properties hold:*

i) The elements of $\overline{E}$ are limits of convergent sequences of elements of E.

ii) $\overline{E}$ is the intersection of all closed subsets that contain E.

iii) $\overline{E}$ is closed.

iv) E is closed if, and only if, $\overline{E} = E$.

v) E is closed if, and only if, it contains the limit of every convergent sequence of elements of E.

Proof. See Royden and Fitzpatrick (2010), Propositions 3 and 6, pages 188-189. □

Now, let $(V, \langle\cdot,\cdot\rangle)$ be an inner product space, and $\|\cdot\|$ the norm induced by $\langle\cdot,\cdot\rangle$. Consider the mapping $d : V \times V \longrightarrow [0, \infty)$ defined by

$$d(f, g) = \|f - g\| \quad \forall f, g \in V$$

It is an immediate consequence of Definition A.66, that d is a distance on V and, therefore, (V, d) is a metric space.

Once we have introduced a distance in an inner product space, we can present the properties of orthogonal complements.

Proposition A.73. *Let $(V, \langle\cdot,\cdot\rangle)$ be an inner product space, and $U \subseteq V$. Then:*

i) $U^{\perp}$ is a closed subspace of V.

ii) If $U \cap U^{\perp} \neq \emptyset$, then $U \cap U^{\perp} = \{0\}$.

iii) If $W \subseteq U$, then $U^{\perp} \subseteq W^{\perp}$.

iv) $\left(\overline{U}\right)^{\perp} = U^{\perp}$

v) $U \subseteq (U^{\perp})^{\perp}$

Proof. See Axler (2020), 8.40. □

As we have seen, inner product spaces satisfy many of the geometric properties of the Euclidean space. To introduce analytical properties we need to add completeness, which will give rise to Hilbert spaces.

Definition A.74. (Hilbert space) A Hilbert space is an inner product space which is complete with respect to the norm induced by the inner product. In other words, a Hilbert space is a Banach space in which the norm is induced by an inner product.

In Hilbert spaces, geometric and topological properties are related, as the following results show.

Proposition A.75. *Let U be a subspace of a Hilbert space H. Then $\overline{U} = (U^{\perp})^{\perp}$.*

Proof. See Axler (2020), 8.41. □

Corollary A.76. *If U is a closed subspace of a Hilbert space, then $U = \left(U^{\perp}\right)^{\perp}$.*

Proof. Immediate from Proposition A.75 and iv) of Proposition A.72. □

The following result is useful when one needs to prove that a certain subspace of a Hilbert space is dense.

Theorem A.77. *Let U be a subspace of a Hilbert space H. Then*

$$\overline{U} = H \Leftrightarrow U^{\perp} = \{0\}$$

i.e., a subspace of a Hilbert space is dense if, and only if, its orthogonal complement contains only the vector zero.

Proof. See Axler (2020), 8.42. □

Before introducing the next result on Hilbert spaces, we need some definitions related to vector spaces.

Definition A.78. (Sum of subspaces) Let A, B be two subspaces of the vector space V. The sum of A and B, denoted by $A + B$, is the set defined by

$$A + B = \{x + y : x \in A, y \in B\}$$

As an immediate consequence of its definition, the sum of two subspaces is also a subspace.

Definition A.79. (Direct sum) If the sum of the subspaces A and B, $A + B$, is such that $A \cap B = \{0\}$, then we write $A + B$ as $A \oplus B$, and call it direct sum of A and B.

Proposition A.80. *The elements of $A \oplus B$ can be uniquely written as the sum of a vector of A plus a vector of B.*

Proof. Let $f \in A \oplus B$, and suppose that $f = g_1 + h_1 = g_2 + h_2$, with $g_1, g_2 \in A$ and $h_1, h_2 \in B$. Then $g_1 - g_2 = h_2 - h_1$ belongs to $A \cap B$, because A and B are subspaces. Then, $g_1 - g_2 = h_2 - h_1 = 0$ and $g_1 = g_2$, $h_1 = h_2$, which concludes the proof. □

Theorem A.81. *If U is a closed subspace of a Hilbert space V, then $V = U \oplus U^{\perp}$.*

Proof. See Axler (2020), 8.43. □

For the next important result, known as Riesz Representation Theorem, we also need some more definitions and preliminary results from Functional Analysis. In the first one, we will use the same notation $||\cdot||$ for different norms. The argument of the norm will determine its meaning.

Definition A.82. (Bounded linear map) Let U and V be normed vector spaces, and $T : U \to V$ a linear map. Define $||T||$ by

$$||T|| = \sup \{||T(f)|| : f \in U, \ ||f|| \leq 1\} \tag{A.6}$$

We say that T is bounded if $||T|| < \infty$, and denote by $\mathcal{B}(U, V)$ the set of bounded linear maps from U to V.

Proposition A.83. *The mapping* $\|\cdot\| : \mathcal{B}(U,V) \to [0,\infty)$ *defined by (A.6) is a norm on* $\mathcal{B}(U,V)$, *and then,* $\mathcal{B}(U,V)$ *is a normed vector space.*

Proof. See Axler (2020), 6.46. □

Definition A.84. (Linear functional) A linear functional on a vector space V is a linear map from V to $\mathbb{R}$.

The inner product of a Hilbert space allows us to define bounded linear functionals associated to each element of the Hilbert space.

Proposition A.85. *Let* H *be a Hilbert space and* $h \in H$. *Then, the mapping* φ_h : $H \to \mathbb{R}$ *defined by*

$$\varphi_h(f) = \langle f, h\rangle \quad \forall f \in H \tag{A.7}$$

is a bounded linear functional.

Proof. φ_h is linear by the properties of the inner product, and it is bounded because by the Cauchy-Schwarz inequality we have

$$\|\varphi_h(f)\| = |\langle f, h\rangle| \le \|f\|\,\|h\| < \infty$$

□

The Riesz Representation Theorem states that every bounded linear functional on a Hilbert space is of the form (A.7).

Theorem A.86. ***(Riesz Representation Theorem)*** *Let* φ *be a bounded linear functional on a Hilbert space* H. *Then, there exists a unique* $h \in H$ *such that*

$$\varphi(f) = \langle f, h\rangle \quad \forall f \in H$$

Furthermore, $\|\varphi\| = \|h\|$.

Proof. See Axler (2020), 8.47. □

We have seen that $L^p(\mu)$ spaces are Banach spaces for $1 \le p \le \infty$ (see Theorem A.56). Therefore, these are our main candidates for being Hilbert spaces. The only property that we have to check is that the p-norm comes from an inner product. Unfortunately this property is satisfied only in the case $p = 2$.

Theorem A.87. $L^2(\mu)$ *is a Hilbert space with the inner product*

$$\langle f, g\rangle = \int_X f(x)\, g(x)\, d\mu(x) \quad \forall f, g \in L^2(\mu) \tag{A.8}$$

Proof. Applying Definition A.44 with $p = 2$, it is immediate to check that the parallelogram identity holds. Then, by Theorem A.61, $L^2(\mu)$ is an inner product space and, by Theorem A.56, it is also a Hilbert space. Finally, by the polarization identity (A.4), we find expression (A.8). □

A.9 The Radon-Nikodym Theorem

To conclude this appendix, we state the Radon-Nikodym Theorem, which has important implications, especially in a probabilistic framework, since it plays a fundamental role in Girsanov's Theorem.

Definition A.88. (Absolutely continuous measure, equivalent measures) Let μ, ν be two measures defined in the same measurable space $(X, \mathcal{F})$. We say that the measure ν is absolutely continuous with respect to μ, denoted by $\nu << \mu$, if $\nu(A) = 0$ for every $A \in \mathcal{F}$ with $\mu(A) = 0$, i.e., if the null sets of μ are also null sets of ν. If $\nu << \mu$ and $\mu << \nu$, then we say that μ and ν are equivalent measures, denoted by $\mu \sim \nu$.

Proposition A.89. *Let $(X, \mathcal{F}, \mu)$ be a measure space and let f be a $\mathcal{F}$-measurable and non-negative function on X. Define the mapping $\nu : \mathcal{F} \to [0, \infty]$ by*

$$\nu(A) = \int_A f(x)\, d\mu(x) \quad \forall A \in \mathcal{F}$$

Then, ν is a measure on $(X, \mathcal{F})$ that satisfies that if $E \in \mathcal{F}$ and $\mu(E) = 0$, then $\nu(E) = 0$.

Proof. For the proof that ν is a measure, see Cohn (2013), Example 2.4.3. For the last part, we have that

$$\nu(E) = \int_E f(x)\, d\mu(x) = \int_X \mathbf{1}_E(x) f(x)\, d\mu(x) = 0$$

because $\mathbf{1}_E = 0$ μ-a.e. □

The Radon-Nikodym Theorem states that the converse of the previous proposition is true, under some assumptions on the measures μ, ν.

Theorem A.90. ***(Radon-Nikodym)*** *Let $(X, \mathcal{F})$ be a measurable space, and μ, ν two σ-finite measures on $(X, \mathcal{F})$, with $\nu << \mu$. Then, there exists a non-negative $\mathcal{F}$-measurable function f on X such that*

$$\nu(A) = \int_A f(x)\, d\mu(x) \quad \forall A \in \mathcal{F} \tag{A.9}$$

The function f is unique in the sense that if g is any non-negative measurable function on X satisfying (A.9), then $g = f$ μ-a.e.

Proof. See Royden and Fitzpatrick (2010), page 382. □

Definition A.91. (Radon-Nikodym derivative) The function f in Theorem A.90 is called the Radon-Nikodym derivative of ν with respect to μ, and is denoted by $\frac{d\nu}{d\mu}$.

Appendix B

Probability Theory

B.1 Definitions and Preliminary Results

As we will see shortly, a probability measure is merely a particular type of measure. Therefore, in this appendix, we will make use of many of the definitions and results from Appendix A, and will only change names, and interpretations, when necessary.

Let Ω be a set, and $\mathcal{F}$ a σ-algebra on Ω.

Definition B.1. (Probability measure, probability space, sample space) A probability measure $\mathbb{P}$ on a measurable space $(\Omega, \mathcal{F})$ is a measure on $(\Omega, \mathcal{F})$ satisfying $\mathbb{P}(\Omega) = 1$. The triple $(\Omega, \mathcal{F}, \mathbb{P})$ is known as probability space, and Ω is called the sample space.

Definition B.2. (Events, probability of an event) If $(\Omega, \mathcal{F}, \mathbb{P})$ is a probability space, the sets $A \in \mathcal{F}$ are known as events, and $\mathbb{P}(A)$ is called the probability of A. The elements of Ω are typically denoted by ω.

Definition B.3. (Almost surely) Let $(\Omega, \mathcal{F}, \mathbb{P})$ be a probability space. We say that an event $A \in \mathcal{F}$ happens almost surely (a.s.) if $\mathbb{P}(A) = 1$.

Definition B.4. (Random variable) An $\mathcal{F}$-measurable function from Ω to $\mathbb{R}$ is called a random variable on $(\Omega, \mathcal{F})$.

Definition B.5. (Expectation) If X is a random variable on $(\Omega, \mathcal{F})$, then the integral of X with respect to the probability measure $\mathbb{P}$ is called the expectation of X, and it is denoted by $\mathbb{E}^{\mathbb{P}}[X]$, or simply $\mathbb{E}[X]$ when there is no other probability measure:

$$\mathbb{E}[X] = \int_{\Omega} X(\omega)\, d\mathbb{P}(\omega)$$

The same possibilities that we had for the integral of a measurable function (see Definitions A.16-A.17 and subsequent comment), apply to the expectation of a random variable. Moreover, all the properties and results related to integrals with respect to a measure hold directly for expectations.

By the following result, we can obtain the probabillity of an event as a particular expectation.

Proposition B.6. *If $A \in \mathcal{F}$, then $\mathbb{P}(A) = \mathbb{E}[\mathbf{1}_A]$.*

Proof. Immediate from Proposition A.13. □

From now on, we will omit the variable ω in most situations. For example, the set $\{\omega \in \Omega : X(\omega) \in U\}$ will be denoted as $\{X \in U\}$. Sometimes, we will even omit the curly braces, and simply write $X \in U$.

Proposition B.7. *Let $(\Omega, \mathcal{F}, \mathbb{P})$ be a probability space, and X a random variable on $(\Omega, \mathcal{F})$. Then, the mapping $\mathbb{P}_X : \mathcal{B}(\mathbb{R}) \to [0,1]$ defined by*

$$\mathbb{P}_X(B) = \mathbb{P}(X \in B) = \mathbb{P}\left[X^{-1}(B)\right]$$

is a probability on $(\mathbb{R}, \mathcal{B}(\mathbb{R}))$. Moreover, if $g : \mathbb{R} \to \mathbb{R}$ is a Borel measurable function, then

$$g(X) \in L^1(\Omega, \mathcal{F}, \mathbb{P}) \Leftrightarrow g \in L^1(\Omega, \mathcal{B}(\mathbb{R}), \mathbb{P}_X)$$

If these functions are integrable, then

$$\int_\Omega g[X(\omega)]\, d\mathbb{P}(\omega) = \int_\mathbb{R} g(x)\, d\mathbb{P}_X(x)$$

Proof. It suffices to realize that $\mathbb{P}_X = \mathbb{P} \circ \mathbb{X}^{-1}$ and apply Proposition A.27. □

The importance of the previous result lies the fact that it allows to transform integrals on Ω, which is generally unknown, into integrals on $\mathbb{R}$, provided the measure $\mathbb{P}_X$ is known.

In terms of expectations, we have the following result.

Proposition B.8. *Let X be a random variable and $g : \mathbb{R} \to \mathbb{R}$ be a measurable function, with $g(X) \in L^1(\Omega, \mathcal{F}, \mathbb{P})$, or equivalently $g \in L^1(\Omega, \mathcal{B}(\mathbb{R}), \mathbb{P}_X)$. Then*

$$\mathbb{E}[g(X)] = \int_\Omega g[X(\omega)]\, d\mathbb{P}(\omega) = \int_\mathbb{R} g(x)\, d\mathbb{P}_X(x)$$

Proof. A direct consequence of Proposition B.7. □

Corollary B.9. *If $X \in L^1(\Omega, \mathcal{F}, \mathbb{P})$, then*

$$\mathbb{E}[X] = \int_\Omega X(\omega)\, d\mathbb{P}(\omega) = \int_\mathbb{R} x d\mathbb{P}_X(x)$$

The same result holds if $i \in L^1(\Omega, \mathcal{B}(\mathbb{R}), \mathbb{P}_X)$, where $i : \mathbb{R} \to \mathbb{R}$ is the identity function given by $i(x) = x$.

Proof. It is a direct consequence of Proposition B.8. □

Definition B.10. (Probability distribution, distribution function) Let $(\Omega, \mathcal{F}, \mathbb{P})$ be a probability space, and X a random variable on $(\Omega, \mathcal{F})$. The probabililty $\mathbb{P}_X : \mathcal{B}(\mathbb{R}) \to [0,1]$ defined in Proposition B.7 is called the probability distribution of X. The function $F_X : \mathbb{R} \to [0,1]$ defined by

$$F_X(x) = \mathbb{P}_X((-\infty, x]) = \mathbb{P}(X \leq x) \tag{B.1}$$

is called the distribution function of X.

Proposition B.11. *The distribution function, F_X, of a random variable X, satisfies the following properties:*

i) F_X *is increasing.*

ii) F_X *is right continuous, i.e.,* $\lim_{x \to s^+} F_X(x) = F_X(s) \quad \forall s \in \mathbb{R}$

iii) $\lim_{x \to +\infty} F_X(x) = 1$

iv) $\lim_{x \to -\infty} F_X(x) = 0$

Proof. See Axler (2020), 12.29. □

The converse of the previous proposition is also true.

Proposition B.12. *Let $F : \mathbb{R} \to [0, 1]$ be a function satisfying i)-iv) of Proposition B.11. Then, there exists a random variable X such that F is the distribution function of X.*

Proof. See Axler (2020), 12.29. □

For most random variables relevant to applications, obtaining probabilities and expectations can be further simplified by means of density functions.

Definition B.13. (Density function) Let X be a random variable, and F_X its associated distribution function. If there exists a non-negative and Lebesgue integrable function f_X, such that

$$\mathbb{P}_X(B) = \int_B f_X(x)\, dx \quad \forall B \in \mathcal{B}(\mathbb{R}) \tag{B.2}$$

we say that f_X is the density function of X.

Remark B.14. By its definition, the density function of a random variable, if it exists, is unique up to changes on sets of Lebesgue measure zero. Moreover, we can write

$$\mathbb{P}_X(B) = \mathbb{E}^{\mathbb{P}_X}[\mathbf{1}_B] = \int_{\mathbb{R}} \mathbf{1}_B d\mathbb{P}_X(x) = \int_B d\mathbb{P}_X(x) \quad \forall B \in \mathcal{B}(\mathbb{R})$$

If X has a density f_X, comparing the previous expression with (B.2), we conclude that $d\mathbb{P}_X(x) = f_X(x)\, dx$.

The density function allows us to write the distribution function as an indefinite integral.

Proposition B.15. *Let X be a random variable with density f_X, and let F_X be its distribution function. Then*

$$F_X(x) = \int_{-\infty}^{x} f_X(y)\, dy \quad \forall x \in \mathbb{R} \tag{B.3}$$

Proof. An immediate consequence of expressions (B.1) and (B.2). □

The following proposition states some properties of density functions.

Proposition B.16. *Let X be a random variable with distribucion function F_X and density function f_X. Then, the following properties hold:*

i) $f_X(x) \geq 0$ *for almost every* $x \in \mathbb{R}$

ii) F_X *is differentiable Lebesgue almost everywhere, and* $F'_X(x) = f_X(x)$ *for almost every* $x \in \mathbb{R}$.

iii) $\int_{-\infty}^{+\infty} f_X(x)\,dx = 1$

iv) $\mathbb{P}(X \leq x) = \int_{-\infty}^{x} f_X(y)\,dy$

Proof. Property i) is immediate from Definition B.13. For ii), see Axler (2020), 4.19. Expression (B.3) and iii) of Proposition B.11 lead to iii). Finally, iv) is just a rewriting of expression (B.3). □

Proposition B.8 and Corollary B.9 can be rewritten with density functions as follows.

Proposition B.17. *Let* X *be a random variable with density* f_X*, and let* $g : \mathbb{R} \to \mathbb{R}$ *be a Borel measurable function, with* $g(X) \in L^1(\Omega, \mathcal{F}, \mathbb{P})$*, or equivalently* $g \in L^1(\mathbb{R}, \mathcal{B}(\mathbb{R}), \mathbb{P}_X)$*. Then*

$$\mathbb{E}[g(X)] = \int_{\mathbb{R}} g(x) f_X(x)\,dx$$

Proof. It suffices to use $d\mathbb{P}_X = f_X dx$ in Proposition B.8. For a more detailed proof, see Shreve (2004), Theorem 1.5.2. □

Corollary B.18. *If* $X \in L^1(\Omega, \mathcal{F}, \mathbb{P})$ *has density* f_X*, then*

$$\mathbb{E}[X] = \int x f_X(x)\,dx$$

The same result holds if $i \in L^1(\mathbb{R}, \mathcal{B}(\mathbb{R}), \mathbb{P}_X)$*, where* $i : \mathbb{R} \to \mathbb{R}$ *is the identity function,* $i(x) = x$.

Proof. It is a direct consequence of Proposition B.17. □

The following definitions are of special importance in the applications of Probability Theory.

Definition B.19. (Moment of a random variable) If X is a random variable, and $k \in \mathbb{N}$, its k-th moment is defined as $\mathbb{E}\left[X^k\right]$.

Definition B.20. (Variance) Let X be a random variable. The variance of X, $\mathrm{Var}(X)$, is defined as the second moment of $X - \mathbb{E}(X)$:

$$\mathrm{Var}(X) = \mathbb{E}\left[(X - \mathbb{E}(X))^2\right]$$

Definition B.21. (Covariance) Let X, Y be two random variables. The covariance of X and Y, $\mathrm{Cov}(X, Y)$, is defined by

$$\mathrm{Cov}(X, Y) = \mathbb{E}[(X - \mathbb{E}(X))(Y - \mathbb{E}(Y))]$$

Definition B.22. (Correlation coefficient, uncorrelated random variables) Let X, Y be two random variables. Then, the correlation coefficient of X and Y, $\rho(X, Y)$, is defined as

$$\rho(X, Y) = \frac{\mathrm{Cov}(X, Y)}{\sqrt{\mathrm{Var}(X)}\sqrt{\mathrm{Var}(Y)}}$$

If $\rho(X, Y) = 0$ (or equivalently $\mathrm{Cov}(X, Y) = 0$), we say that X and Y are uncorrelated random variables.

Note that, as in the case of expectations, according to the previous definitions, moments, variance, covariance and correlation coefficient, can have finite or infinite values, or even be undefined.

The following two results state some properties of variance, covariance and the correlation coefficient.

Proposition B.23. *Let X, Y two random variables with $X, Y \in L^2(\Omega, \mathcal{F}, \mathbb{P})$. Then:*

i) $\mathrm{Var}(X) = \mathrm{Cov}(X, X)$

ii) $\mathrm{Var}(X) = \mathbb{E}\left[X^2\right] - (\mathbb{E}[X])^2$

iii) $\mathrm{Cov}(X, Y) = \mathrm{Cov}(Y, X) = \mathbb{E}[XY] - \mathbb{E}[X]\,\mathbb{E}[Y]$

iv) $\mathrm{Cov}\left(\sum_{i=1}^k a_i X_i, \sum_{j=1}^l b_j Y_j\right) = \sum_{i=1}^k \sum_{j=1}^l a_i b_j \mathrm{Cov}\left(X_i, Y_j\right)$

v) $\mathrm{Var}\left(\sum_{i=1}^k a_i X_i\right) = \sum_{i=1}^k a_i^2 \mathrm{Var}(X_i) + 2\sum_{1 \le i < j \le k} a_i a_j \mathrm{Cov}\left(X_i, X_j\right)$

vi) $\mathrm{Cov}(aX, bY) = ab\mathrm{Cov}(X, Y)$

vii) $\mathrm{Var}(aX) = a^2 \mathrm{Var}(X)$

Proof. i), ii) and iii) are direct consequences of the definitions. For iv) and v), see Resnick (2014), pages 129-130. Finally, properties vi) and vii) are immediate from iv) and v). □

Proposition B.24. *Let X, Y be two random variables with $X, Y \in L^2(\Omega, \mathcal{F}, \mathbb{P})$, and $\mathrm{Var}(X), \mathrm{Var}(Y) > 0$. Then:*

i) $-1 \le \rho(X, Y) \le 1$

ii) $\rho(X, X) = 1$

Proof. For property i), defining the random variables $U = X - \mathbb{E}[X]$ and $V = Y - \mathbb{E}[Y]$, we have that $\mathrm{Cov}(X, Y) = \mathbb{E}[UV]$ and $\mathrm{Var}(X) = \mathbb{E}\left[U^2\right]$, $\mathrm{Var}(Y) = \mathbb{E}\left[V^2\right]$. On the other hand, by the Cauchy-Schwarz inequality in the Hilbert space $L^2(\Omega, \mathcal{F}, \mathbb{P})$, we have that

$$|\mathbb{E}[UV]| \le \sqrt{\mathbb{E}\left[U^2\right]}\sqrt{\mathbb{E}\left[V^2\right]}$$

and then

$$|\rho(X, Y)| = \left|\frac{\mathrm{Cov}(X, Y)}{\sqrt{\mathrm{Var}(X)}\sqrt{\mathrm{Var}(Y)}}\right| = \left|\frac{\mathbb{E}[UV]}{\sqrt{\mathbb{E}\left[U^2\right]}\sqrt{\mathbb{E}\left[V^2\right]}}\right| \le 1$$

Property ii) is immediate from the definition of the correlation coefficient and i) of Proposition B.23. □

The following result contains some useful inequalities.

Proposition B.25. *Let X be an integrable random variable. Then, the following inequalities hold:*

i) (Markov inequality)

$$\mathbb{P}\left[|X| \geq \alpha\right] \leq \frac{1}{\alpha^k}\mathbb{E}\left[|X|^k\right] \quad \forall \alpha, k > 0$$

ii) (Chebyshev inequality) If $\mathrm{Var}(X)$ *is well defined and finite, then*

$$\mathbb{P}\left[|X - \mathbb{E}[X]| \geq \alpha\right] \leq \frac{1}{\alpha^2}\mathrm{Var}(X) \quad \forall \alpha > 0$$

iii) (Jensen inequality) If φ *is convex in any interval containing the range of* X*, and* $\varphi(X)$ *is integrable, then*

$$\varphi\left(\mathbb{E}[X]\right) \leq \mathbb{E}\left[\varphi(X)\right]$$

Proof. See Billingsley (1995), page 80. □

B.2 Independence

Considering a probability measure into a measurable space allows us to discuss the independence of events. To gain an intuitive understanding of this concept, it is helpful to begin by defining conditional probability.

Let $(\Omega, \mathcal{F}, \mathbb{P})$ be a probability space and suppose a random experiment has been performed, yielding an outcome $\omega \in B$, with $B \in \mathcal{F}$. Knowing this result can alter the probability of another event since, in a sense, it is as if the probability space has been reduced from Ω to B. The probability of an event obtained knowing that another event B has occurred is called probability conditional on B. Its formal definition is as follows.

Definition B.26. (Conditional probability given an event) Let B be an event with $\mathbb{P}(B) > 0$. The mapping $\mathbb{P}(\cdot|B) : \mathcal{F} \to [0, 1]$ given by

$$\mathbb{P}(A|B) = \frac{\mathbb{P}(A \cap B)}{\mathbb{P}(B)} \quad \forall A \in \mathcal{F}$$

is called the conditional probability of A given B.

Proposition B.27. $\mathbb{P}(\cdot|B)$ *is a probability measure on* $(\Omega, \mathcal{F})$.

Proof. First, $\mathbb{P}(\cdot|B)$ is a non-negative set function because

$$\mathbb{P}(A|B) = \frac{\mathbb{P}(A \cap B)}{\mathbb{P}(B)} \geq 0$$

since $\mathbb{P}$ is a measure on $(\Omega, \mathcal{F})$. Moreover,

$$\mathbb{P}(\varnothing|B) = \frac{\mathbb{P}(\varnothing \cap B)}{\mathbb{P}(B)} = \frac{\mathbb{P}(\varnothing)}{\mathbb{P}(B)} = 0$$

and

$$\mathbb{P}(\Omega|B) = \frac{\mathbb{P}(\Omega \cap B)}{\mathbb{P}(B)} = \frac{\mathbb{P}(B)}{\mathbb{P}(B)} = 1$$

Finally, if $A_n \in \mathcal{F}\ \forall n \in \mathbb{N}$, with $A_i \cap A_j = \varnothing$ for $i \neq j$, then

$$\begin{aligned}\mathbb{P}\left(\bigcup_{n=1}^{\infty} A_n|B\right) &= \frac{\mathbb{P}\left(\bigcup_{n=1}^{\infty} A_n \cap B\right)}{\mathbb{P}(B)} = \frac{\mathbb{P}\left[\bigcup_{n=1}^{\infty}(A_n \cap B)\right]}{\mathbb{P}(B)} \\ &= \frac{\sum_{n=1}^{\infty} \mathbb{P}(A_n \cap B)}{\mathbb{P}(B)} = \sum_{n=1}^{\infty} \frac{\mathbb{P}(A_n \cap B)}{\mathbb{P}(B)} \\ &= \sum_{n=1}^{\infty} \mathbb{P}(A_n|B)\end{aligned}$$

□

Definition B.28. (Independent events) Let A, B be two events. We say that A and B are independent if

$$\mathbb{P}(A \cap B) = \mathbb{P}(A)\,\mathbb{P}(B)$$

Let $\{A_i\}_{i=1}^n$ be a finite family of events. We say that $A_1, \dots, A_n$ are independent if

$$\mathbb{P}\left(\bigcap_{i \in I} A_i\right) = \prod_{i \in I} \mathbb{P}(A_i) \quad \forall I \subset \{1, \dots, n\}$$

Remark B.29. If $A_1, \dots, A_n$ are independent events, then the events A_i and A_j, with $i \neq j$, are independent. The converse is not true, in general, for $n > 2$. Pairwise independence does not imply mutual independence. Furthermore, independence is a property that depends on the probability measure.

Proposition B.30. *Let A, B be two events with $\mathbb{P}(B) > 0$. Then, A and B are independent if, and only if, $\mathbb{P}(A|B) = \mathbb{P}(A)$.*

Proof. Immediate from Definitions B.26 and B.28. □

Now, we can provide the intuition behind Definition B.28. According to the previous proposition, two events are independent when the probability of one of them is not affected by the knowledge that the other event has occurred.[1]

The concept of independence can be extended to other mathematical objects.

Definition B.31. (σ-algebra generated by a random variable) The σ-algebra generated by the random variable X, $\sigma(X)$, is the smallest σ-algebra with respect to which X is measurable.

Definition B.32. (Sub-σ-algebra) Let $(\Omega, \mathcal{F})$ be a measurable space. A sub-σ-algebra of $\mathcal{F}$ is a family $\mathcal{G} \subset \mathcal{F}$ such that $\mathcal{G}$ is itself a σ-algebra on Ω.

[1]We say that an event A has occurred if, when performing a random experiment, an outcome $\omega \in A$ is obtained.

Definition B.33. (Independence of σ-algebras and random variables) Let $(\Omega, \mathcal{F}, \mathbb{P})$ be a probability space, and $\mathcal{G}, \mathcal{H}$ sub-σ-algebras of $\mathcal{F}$. We say that $\mathcal{G}$ and $\mathcal{H}$ are independent if

$$\mathbb{P}(A \cap B) = \mathbb{P}(A)\,\mathbb{P}(B) \quad \forall A \in \mathcal{G} \ \forall B \in \mathcal{H}$$

If $\mathcal{G}_1, \ldots, \mathcal{G}_n$ are sub-σ-algebras of $\mathcal{F}$, we say that $\mathcal{G}_1, \ldots, \mathcal{G}_n$ are independent if

$$\mathbb{P}(A_1 \cap \cdots \cap A_n) = \prod_{i=1}^{n} \mathbb{P}(A_i) \quad \forall A_i \in \mathcal{G}_i \ \ i = 1, \ldots, n$$

The random variables X and Y are independent if the σ-algebras generated by X and Y, $\sigma(X)$ and $\sigma(Y)$, are independent. The random variables $X_1, \ldots, X_n$ are independent if the σ-algebras $\sigma(X_1), \ldots, \sigma(X_n)$ are independent. The random variable X is independent of the σ-algebra $\mathcal{G}$ if $\sigma(X)$ and $\mathcal{G}$ are independent.

Remark B.34. As in the previous remark, pairwise independence of sub-σ-algebras and random variables does not imply mutual independence.

Starting with two independent random variables, it is possible to construct other independent random variables by applying measurable functions.

Proposition B.35. *Let X, Y be independent random variables, and f, g two Borel measurable functions on $\mathbb{R}$. Then, $f(X)$ and $g(Y)$ are independent random variables.*

Proof. See Shreve (2004), Theorem 2.2.5. □

The independence of random variabes simplify the work with expectations, variances and covariances.

Proposition B.36. *Let X, Y be two integrable and independent random variables. Then:*

i) XY is integrable and

$$\mathbb{E}[XY] = \mathbb{E}[X]\,\mathbb{E}[Y]$$

Moreover, if $X_1, \ldots, X_n$ are independent and integrable random variables, then $X_1, \cdots, X_n$ is integrable and

$$\mathbb{E}[X_1 \cdots X_n] = \mathbb{E}[X_1] \cdots \mathbb{E}[X_n]$$

ii) X and Y are uncorrelated: $\operatorname{Cov}(X, Y) = 0$. *If the correlation coefficient is defined, then $\rho(X, Y) = 0$.*

iii) $\operatorname{Var}(aX + bY) = a^2 \operatorname{Var}(X) + b^2 \operatorname{Var}(Y) \quad \forall a, b \in \mathbb{R}$

Proof. For property i), see Billingsley (1995), page 277. Property ii) is immediate from i), and iii) of Proposition B.23. Finally, property iii) is also immediate from ii) and v) of Proposition B.23. □

It is important to note that although the independence of two random variables implies that these variables are uncorrelated, the converse is not true in general.[2]

[2] See Shreve (2004), Example 2.2.10, for an example of two uncorrelated random variables which are not independent.

B.3 Moment Generating and Characteristic Functions

The Laplace and Fourier transforms of a random variable characterize the distribution of that variable, and play a key role in Probability Theory.

Definition B.37. (Moment generating function) Let X be a random variable. The moment generating function of X is the function $M_X : D \subset \mathbb{R} \to (0, \infty]$, defined on the domain $D = \{u \in \mathbb{R} : \mathbb{E}\left[e^{uX}\right] < \infty\}$ by

$$M_X(u) = \mathbb{E}\left[e^{uX}\right]$$

From Definition B.37, it is immediate to obtain $M_X(u) > 0 \ \forall u \in D$, that $0 \in D$ with $M_X(0) = 1$, and that if $Y = aX + b$, with $a, b \in \mathbb{R}$, then $M_Y(u) = e^{bu} M_X(au)$.

The most important properties of the moment generating function require an additional assumption.

Proposition B.38. *Let X, Y be two random variables with $M_X(u), M_Y(u)$ finite on an interval $(-s, s)$, with $s > 0$. Then:*

i) $M_X^{(n)}(0) = \mathbb{E}[X^n] \ \forall n \in \mathbb{N}$

ii) If $M_X(u) = M_Y(u)$ on $(-s, s)$, then $F_X(x) = F_Y(x) \ \forall x \in \mathbb{R}$

iii) If $X_1, \cdots, X_n$ are independent and their moment generating functions are finite on $(-s, s)$, then $M_{X_1+\cdots+X_n}(u) = M_{X_1}(u) \cdots M_{X_n}(u)$ in $(-s, s)$.

Proof. See Billingsley (1995), pages 278, 280 and 390. □

The first statement of the previous proposition justifies the name of moment generating function. The second one tells us that the moment generating function characterizes the distribution of a random variable. Nevertheless, its validity depends on the finiteness of the moment generating function on an interval containing zero, which is not guaranteed in all cases.[3] This is not what happens with characteristic functions that, as we will see in what follows, always exist and also determine uniquely distribution functions.

Definition B.39. (Characteristic function) Let X be a random variable. The characteristic function of X is the function $\varphi_X : \mathbb{R} \to \mathbb{C}$ defined by

$$\varphi_X(u) = \mathbb{E}\left[e^{iuX}\right]$$

where $i = \sqrt{-1}$.

Note that

$$|\varphi_X(u)| = \left|\mathbb{E}\left[e^{iuX}\right]\right| \leq \mathbb{E}\left[\left|e^{iuX}\right|\right] = 1 \quad \forall u \in \mathbb{R}$$

and then, the characteristic function is always well defined and finite.

The following proposition states some properties of the characteristic function.

Proposition B.40. *Let X, Y be two random variables and φ_X, φ_Y their respective characteristic functions. Then:*

[3] Some texts state that the moment generating function exists only if it is finite on $(-s, s)$ with $s > 0$.

i) $\varphi_X(0) = 1$

ii) $\varphi_X(u) = M_X(iu)\ \forall u \in \mathbb{R}$

iii) If $Y = aX + b$, *with* $a, b \in \mathbb{R}$, *then* $\varphi_Y(u) = e^{ibu}\varphi_X(au)$

iv) If $X_1, \cdots, X_n$ *are independent, then*

$$\varphi_{X_1+\cdots+X_n}(u) = \varphi_{X_1}(u)\cdots\varphi_{X_n}(u) \quad \forall u \in \mathbb{R}$$

v) φ_X *is uniformly continuous on* $\mathbb{R}$

vi) If $k \in \mathbb{N}$ *and* $\mathbb{E}\left[|X|^k\right] < \infty$, *then* $\varphi_X^{(k)}(0) = i^k\mathbb{E}\left[X^k\right]$

vii) If $\varphi_X(u) = \varphi_Y(u)\ \forall u \in \mathbb{R}$, *then* $F_X(x) = F_Y(x)\ \forall x \in \mathbb{R}$

viii) If φ_X *is integrable, then* X *has a bounded continuous density* f_X *given by*

$$f_X(x) = \frac{1}{2\pi}\int_{\mathbb{R}} e^{-iux}\varphi_X(u)\,du$$

Proof. Properties i)-iii) are immediate from the definition of the characteristic function. The proof of iv) follows from the definition and i) of Proposition B.36. For properties v)-viii), see Resnick (2014), Sections 9.2, 9.4 and 9.5. □

The property vi) of the previous proposition tells us that characteristic functions can also be used as moment generating functions. Moreover, by vii), they characterize distributions without the need for additional assumptions. Property viii) allows to recover density functions from characteristic functions, and it is known as *Fourier inversion formula*.

B.4 The Normal Distribution

Before advancing further in the study of Probability Theory, let us focus on the most important probability distribution: the normal distribution.

Definition B.41. (Normal distribution, standard normal distribution) Let X be a random variable. We say that X has a normal or Gaussian distribution with parameters $\mu \in \mathbb{R}$ and $\sigma \in \mathbb{R}^+$, denoted by $X \sim N\left(\mu, \sigma^2\right)$, if its density function has the form

$$f_X(x) = \frac{1}{\sqrt{2\pi}\sigma} e^{-\dfrac{(x-\mu)^2}{2\sigma^2}}$$

The distribution $N(0, 1)$ is called standard normal distribution, and its density function is usually denoted by $\phi(x)$.

By its definition, and taking into account that $\int_{-\infty}^{+\infty} e^{-\frac{(x-\mu)^2}{2\sigma^2}}\,dx = \sqrt{2\pi}\sigma$, it is immediate to check that the density function of the normal distribution satisfies properties i) and iii) of Proposition B.16.

Remark B.42. Sometimes it is useful to consider the random variable $X = \mu \in \mathbb{R}$ as having a normal distribution with $\sigma = 0$. This distribution, $N(\mu, 0)$, is called *degenerate normal distribution*, and it has no density with respect to the Lebesgue measure.

Proposition B.43. *Let X be a random variable with distribution $N\left(\mu, \sigma^2\right)$. Then:*

i) $\mathbb{E}[X] = \mu$, $\operatorname{Var}(X) = \sigma^2$

ii) The distribution function of X is given by

$$F_X(x) = \frac{1}{\sqrt{2\pi}\sigma} \int_{-\infty}^{x} e^{-\frac{(y-\mu)^2}{2\sigma^2}} dy$$

When $X \sim N(0, 1)$, we usually write $F_X(x)$ as $\Phi(x)$.

iii) $M_X(u) = \exp\left(\mu u + \frac{1}{2}\sigma^2 u^2\right)$, $u \in \mathbb{R}$

iv) $\varphi_X(u) = \exp\left(i\mu u - \frac{1}{2}\sigma^2 u^2\right)$, $u \in \mathbb{R}$

Proof. Properties i)-iv) follow by direct calculation from the definitions. □

By property i) of the previous proposition, the parameters μ and σ^2 are, respectively, the expectation and the variance of a random variable with distribution $N\left(\mu, \sigma^2\right)$. Moreover, by ii), these two parameters characterize completely the distribution of a normal random variable.

B.5 Convergence

In probability theory, there are several concepts of convergence for random variables, some of which are used in this book. We dedicate this section to define these concepts, studying their properties and the relationships between them.

Definition B.44. (Almost sure convergence) A sequence of random variables $\{X_n\}_{n=1}^{\infty}$, is said to converge almost surely (a.s.) to the random variable X, denoted $X_n \xrightarrow{a.s} X$, if $\lim_{n\to\infty} X_n(\omega) = X(\omega)$ for almost every $\omega \in \Omega$.

Definition B.45. (Convergence in probability) The sequence $\{X_n\}_{n=1}^{\infty}$ of random variables converges in probability to the random variable X, denoted $X_n \xrightarrow{\mathbb{P}} X$, if for every $\varepsilon > 0$ we have

$$\lim_{n\to\infty} \mathbb{P}\{|X_n - X| > \varepsilon\} = 0$$

Proposition B.46. *If $\{X_n\}_{n=1}^{\infty}$ converges a.s. to X, then it converges in probability to X.*

Proof. See Chung (2001), Theorem 4.1.2. □

Definition B.47. (L^p-convergence) Let p be a fixed positive real number. The sequence of random variables $\{X_n\}_{n=1}^{\infty}$ converges to the random variable X in L^p, denoted $X_n \xrightarrow{L^p} X$, if $\lim_{n\to\infty} \mathbb{E}[|X_n - X|^p] = 0$.

Proposition B.48. *If $\{X_n\}_{n=1}^{\infty}$ converges in L^p to X, then it converges to X in probability.*

Proof. See Chung (2001), Theorem 4.1.4. □

Remark B.49. There is no direct implication between almost sure convergence and L^p convergence (see Chung (2001), Examples 1 and 2, pages 72-73).

The following result is useful in the study of stochastic integration.

Proposition B.50. *Let $\{X_n\}_{n=1}^{\infty}$ be a sequence of random variables, and X another random variable, with $X_n, X \in L^2(\Omega, \mathcal{F}, \mathbb{P})$ $\forall n \in \mathbb{N}$. If $X_n \xrightarrow{L^2} X$ then $\lim_{n\to\infty} \mathbb{E}[X_n] = \mathbb{E}[X]$ and $\lim_{n\to\infty} \mathbb{E}[X_n^2] = \mathbb{E}[X^2]$.*

Proof. By the properties of expectations, we have

$$\lim_{n\to\infty} |\mathbb{E}[X_n] - \mathbb{E}[X]| = \lim_{n\to\infty} |\mathbb{E}[X_n - X]| \le \lim_{n\to\infty} \mathbb{E}[|X_n - X|] \tag{B.4}$$

On the other hand

$$\begin{aligned} \mathbb{E}[|X_n - X|] &= \mathbb{E}[|(X_n - X) \cdot 1|] = \|(X_n - X) \cdot 1\|_1 \le \|(X_n - X)\|_2 \|1\|_2 \\ &= \|(X_n - X)\|_2 \end{aligned}$$

where we have used Hölder's inequality. Since $X_n \xrightarrow{L^2} X$, $\lim_{n\to\infty} \|(X_n - X)\|_2 = 0$, and then $\lim_{n\to\infty} \mathbb{E}[|X_n - X|] = 0$. Replacing this result in (B.4), we obtain $\lim_{n\to\infty} |\mathbb{E}[X_n] - \mathbb{E}[X]|=0$, and therefore $\lim_{n\to\infty} \mathbb{E}[X_n] = \mathbb{E}[X]$.
For the second result, using again Hölder's inequality we have

$$\mathbb{E}\left[\left|X_n^2 - X^2\right|\right] = \mathbb{E}[|(X_n + X)(X_n - X)|] \le \|X_n + X\|_2 \|X_n - X\|_2 \tag{B.5}$$

By the triangle inequality we can write

$$\|X_n + X\|_2 \le \|X_n - X\|_2 + 2\|X\|_2$$

and since $X_n \xrightarrow{L^2} X$, we have that $\|X_n + X\|_2$ is bounded. Taking limits in (B.5), and using that $\lim_{n\to\infty} \|(X_n - X)\|_2 = 0$, we conclude that $\lim_{n\to\infty} \mathbb{E}\left[|X_n^2 - X^2|\right] = 0$, and then $\lim_{n\to\infty} \mathbb{E}[X_n^2] = \mathbb{E}[X^2]$. □

Definition B.51. (Convergence in distribution) We say that the sequence of random variables $\{X_n\}_{n=1}^{\infty}$ converges in distribution to the random variable X, denoted $X_n \xrightarrow{d} X$, if $\lim_{n\to\infty} F_{X_n}(x) = F_X(x)$ for every x at which F_X is continuous.

Proposition B.52. *If $\{X_n\}_{n=1}^{\infty}$ converges to X in probability, then $\{X_n\}_{n=1}^{\infty}$ converges to X in distribution.*

Proof. See Chung (2001), Theorem 4.4.5. □

Below we provide a schematic summary of the relationships between the different types of convergence. The converse implications do not hold in general.

$$X_n \xrightarrow{a.s.} X \implies X_n \xrightarrow{\mathbb{P}} X \implies X_n \xrightarrow{d} X$$

$$X_n \xrightarrow{L^p} X \implies X_n \xrightarrow{\mathbb{P}} X$$

The following result relates convergence in distribution to the pointwise convergence of characteristic functions. It is usually known as *Lévy's continuity theorem.*

Theorem B.53. *Let $\{X_n\}_{n=1}^{\infty}$ be a sequence of random variables, and $\{\varphi_n\}_{n=1}^{\infty}$ the associated sequence of characteristic functions, i.e., $\varphi_n(u) = \mathbb{E}\left[e^{iuX_n}\right]$ $\forall u \in \mathbb{R}$ $\forall n \in \mathbb{N}$. Let X be another random variable, with characteristic function φ, and such that* $\lim_{n\to\infty} \varphi_n(u) = \varphi(u)$ $\forall u \in \mathbb{R}$. *Then, X_n converges to X in distribution.*

Proof. See Billingsley (1995), Theorem 26.3. □

B.6 Conditional Expectations

In this section we will extend the concept of conditional probability studied in Section B.2 to obtaining expectations conditioned on the knowledge of events, random variables and σ-algebras.

Definition B.54. (Conditional expectation with respect to an event) Let X be an integrable random variable, i.e., $X \in L^1(\Omega, \mathcal{F}, \mathbb{P})$, and $B \in \mathcal{F}$ an event, with $\mathbb{P}(B) > 0$. We define the conditional expectation of X with respect to B, $\mathbb{E}[X|B]$, as

$$\mathbb{E}[X|B] = \frac{1}{\mathbb{P}(B)} \int_B X(\omega)\, d\mathbb{P}(\omega) \tag{B.6}$$

Proposition B.55. *The conditional expectation in (B.6) can be rewritten as*

$$\mathbb{E}[X|B] = \int_\Omega X(\omega)\, d\mathbb{P}(\omega|B)$$

Proof. We have that

$$\begin{aligned}\mathbb{E}[X|B] &= \int_\Omega X(\omega)\, d\mathbb{P}(\omega|B) = \int_B X(\omega)\, d\mathbb{P}(\omega|B) + \int_{\Omega-B} X(\omega)\, d\mathbb{P}(\omega|B) \\ &= \frac{1}{\mathbb{P}(B)} \int_B X(\omega)\, d\mathbb{P}(\omega)\end{aligned}$$

where we have used Definition B.26. □

Proposition B.55 allows us to give a more intuitive interpretation to the conditional expectation $\mathbb{E}[X|B]$ as the expectation of the random variable X calculated with respect to the probability measure $\mathbb{P}(\cdot|B)$, namely, the conditional probability given B. Moreover, Definition B.54 is coherent with the usual form of obtaining probabilities from expectations, as the following result states.

Proposition B.56. *Let $A, B \in \mathcal{F}$, with $\mathbb{P}(B) > 0$. Then*

$$\mathbb{E}[\mathbf{1}_A | B] = \mathbb{P}(A | B)$$

Proof. By Proposition B.55 we have

$$\mathbb{E}[\mathbf{1}_A | B] = \int_\Omega \mathbf{1}_A(\omega)\, d\mathbb{P}(\omega | B) = \mathbb{P}(A | B)$$

where in the last step we have used Proposition A.13. □

Definition B.57. (Conditional expectation with respect to a σ-algebra) Let $X \in L^1(\Omega, \mathcal{F}, \mathbb{P})$ and let $\mathcal{G}$ be a sub-σ-algebra of $\mathcal{F}$. The conditional expectation of X with respect to $\mathcal{G}$, $\mathbb{E}[X | \mathcal{G}]$, is a random variable satisfying:

i) $\mathbb{E}[X | \mathcal{G}]$ is $\mathcal{G}$-measurable and integrable.

ii) $\int_G \mathbb{E}[X | \mathcal{G}]\, d\mathbb{P} = \int_G X d\mathbb{P} \ \ \forall G \in \mathcal{G}$

Theorem B.58. *The conditional expectation $\mathbb{E}[X | \mathcal{G}]$ exists and is unique a.s.*

Proof. See Billingsley (1995), page 445. □

The conditional expectation $\mathbb{E}[X | \mathcal{G}]$ should be understood as the best estimate of $\mathbb{E}[X]$ that we can obtain using only the information contained in the σ-algebra $\mathcal{G}$. In other words, the value $\mathbb{E}[X | \mathcal{G}](\omega)$, "is to be interpreted as the expected value of X for someone who knows for each G in $\mathcal{G}$ whether or not it contains the point ω." (Billingsley (1995), page 445).

The previous interpretation of conditional expectation is especially relevant when random variables can evolve over time, giving rise to stochastic processes. In this case, the information available in a σ-algebra can change over time, leading to the concept of filtration.

Definition B.59. (Filtration) A collection $\{\mathcal{F}_t\}_{t \geq 0}$ of sub-σ-algebras of $\mathcal{F}$ is a filtration if $\mathcal{F}_s \subset \mathcal{F}_t$ for all s, t with $s < t$.

Definition B.60. (Usual hypotheses) The filtration $\{\mathcal{F}_t\}_{t \geq 0}$ is said to satisfy the usual hypotheses if:

i) $\mathcal{F}_0$ contains all the $\mathbb{P}$-null sets of $\mathcal{F}$.

ii) $\mathcal{F}_t = \cap_{u > t} \mathcal{F}_u$, for all $t \geq 0$.

Definition B.61. (Filtered probability space) A filtered probability space is a quadruple $(\Omega, \mathcal{F}, \{\mathcal{F}_t\}_{t \geq 0}, \mathbb{P})$, where $(\Omega, \mathcal{F}, \mathbb{P})$ is a probability space and $\{\mathcal{F}_t\}_{t \geq 0}$ a filtration.

The following result contains some useful properties of conditional expectations.

Proposition B.62. *Let X, Y be two integrable random variables, and $\{X_n\}_{n=1}^{\infty}$ a sequence of integrable random variables on the probability space $(\Omega, \mathcal{F}, \mathbb{P})$. Let $\mathcal{G}$ be a sub-σ-algebra of $\mathcal{F}$. Then, the following properties hold.*

i) $\mathbb{E}[X | \{\emptyset, \Omega\}] = \mathbb{E}[X]$ *a.s.*

ii) $\mathbb{E}[X | \mathcal{F}] = X$ *a.s.*

iii) *If* X *is* $\mathcal{G}$*-measurable, then* $\mathbb{E}[X|\mathcal{G}] = X$ *a.s.*

iv) *If* $X = a \in \mathbb{R}$ *a.s., then* $\mathbb{E}[X|\mathcal{G}] = a$ *a.s.*

v) $\mathbb{E}[aX + bY|\mathcal{G}] = a\mathbb{E}[X|\mathcal{G}] + b\mathbb{E}[Y|\mathcal{G}]$ *a.s.* $\forall a, b \in \mathbb{R}$

vi) *If* $X \leq Y$ *a.s., then* $\mathbb{E}[X|\mathcal{G}] \leq \mathbb{E}[Y|\mathcal{G}]$ *a.s.*

vii) $|\mathbb{E}[X|\mathcal{G}]| \leq \mathbb{E}[|X||\mathcal{G}]$ *a.s.*

viii) *(Dominated Convergence Theorem for conditional expectations) If* $\lim_{n\to\infty} X_n = X$ *a.s. and* $|X_n| \leq Y$ *a.s.* $\forall n \in \mathbb{N}$*, with* Y *integrable, then*

$$\lim_{n\to\infty} \mathbb{E}[X_n|\mathcal{G}] = \mathbb{E}[X|\mathcal{G}] \quad a.s.$$

ix) *(Monotone Convergence Theorem for conditional expectations) If* $X_n \geq 0$ *a.s.* $\forall n \in \mathbb{N}$*,* $\{X_n\}_{n=1}^{\infty}$ *is increasing a.s., and* $\lim_{n\to\infty} X_n = X$ *a.s., then* $\{\mathbb{E}[X_n|\mathcal{G}]\}_{n=1}^{\infty}$ *is increasing a.s., and*

$$\lim_{n\to\infty} \mathbb{E}[X_n|\mathcal{G}] = \mathbb{E}[X|\mathcal{G}] \quad a.s.$$

x) *If* X *is* $\mathcal{G}$*-measurable, and* XY *is integrable, then*

$$\mathbb{E}[XY|\mathcal{G}] = X\mathbb{E}[Y|\mathcal{G}] \quad a.s.$$

xi) *(Tower property of conditional expectations) If* $\mathcal{H}$ *is a sub-*σ*-algebra of* $\mathcal{F}$ *such that* $\mathcal{G} \subset \mathcal{H} \subset \mathcal{F}$*, then*

$$\mathbb{E}[\mathbb{E}[X|\mathcal{G}]|\mathcal{H}] = \mathbb{E}[\mathbb{E}[X|\mathcal{H}]|\mathcal{G}] = \mathbb{E}[X|\mathcal{G}] \quad a.s.$$

xii) $\mathbb{E}[\mathbb{E}[X|\mathcal{G}]] = \mathbb{E}[X]$

xiii) *If* X *is independent of* $\mathcal{G}$*, then* $\mathbb{E}[X|\mathcal{G}] = \mathbb{E}[X]$ *a.s.*

xiv) *(Jensen inequality for conditional expectations) If* $\varphi : \mathbb{R} \to \mathbb{R}$ *is a convex function and* $\varphi(X)$ *is integrable, then*

$$\varphi(\mathbb{E}[X|\mathcal{G}]) \leq \mathbb{E}[\varphi(X)|\mathcal{G}] \quad a.s.$$

xv) *Let* $\phi : \mathbb{R} \times \mathbb{R} \to \mathbb{R}$ *be a bounded Borel-measurable function. Let* X *be* $\mathcal{G}$*-measurable, and* Y *independent of* $\mathcal{G}$*. Define* $f_\phi(x) = \mathbb{E}[\phi(x, Y)]$*. Then*

$$\mathbb{E}[\phi(X,Y)|\mathcal{G}] = f_\phi(X) \quad a.s.$$

Proof. See Resnick (2014), Section 10.3, and Billingsley (1995), pages 446-449. □

Using conditional expectations, we can define conditional variances and covariances as follows.

Definition B.63. (Conditional variance and covariance with respect to a σ-algebra) Let $X, Y \in L^2(\Omega, \mathcal{F}, \mathbb{P})$, and let $\mathcal{G}$ be a sub-σ-algebra of $\mathcal{F}$. The conditional variance of X with respect to $\mathcal{G}$, $\operatorname{Var}(X|\mathcal{G})$, and the conditional covariance of X and Y with respect to $\mathcal{G}$, $\operatorname{cov}(X, Y|\mathcal{G})$, are defined by

$$\operatorname{Var}(X|\mathcal{G}) = \mathbb{E}\left[(X - \mathbb{E}[X|\mathcal{G}])^2 |\mathcal{G}\right] = \mathbb{E}\left[X^2 |\mathcal{G}\right] - (\mathbb{E}[X|\mathcal{G}])^2$$

$$\begin{aligned}\operatorname{Cov}(X, Y|\mathcal{G}) &= \mathbb{E}[(X - \mathbb{E}[X|\mathcal{G}])(Y - \mathbb{E}[Y|\mathcal{G}])|\mathcal{G}] \\ &= \mathbb{E}[XY|\mathcal{G}] - (\mathbb{E}[X|\mathcal{G}])(\mathbb{E}[Y|\mathcal{G}])\end{aligned}$$

The relationship between conditional variances and covariances, and their unconditional versions is stated in the following result.

Proposition B.64. *If $X, Y \in L^2(\Omega, \mathcal{F}, \mathbb{P})$, and $\mathcal{G}$ is a sub-σ-algebra of $\mathcal{F}$, then:*

i) $\operatorname{Var}(X) = \operatorname{Var}(\mathbb{E}[X|\mathcal{G}]) + \mathbb{E}[\operatorname{Var}(X|\mathcal{G})]$

ii) $\operatorname{Cov}(X, Y) = \operatorname{Cov}(\mathbb{E}[X|\mathcal{G}], \mathbb{E}[Y|\mathcal{G}]) + \mathbb{E}[\operatorname{Cov}(X, Y|\mathcal{G})]$

Proof. See Kallenberg (2021), Lemma 8.2. □

Conditional probabilities with respect to σ-algebras can be naturally defined from the corresponding conditional expectations.

Definition B.65. (Conditional probability given a σ-algebra) Let $\mathcal{G}$ be a sub-σ-algebra of $\mathcal{F}$, and $A \in \mathcal{F}$. The conditional probability of the event A given $\mathcal{G}$, $\mathbb{P}(A|\mathcal{G})$, is the random variable determined by

$$\mathbb{P}(A|\mathcal{G}) = \mathbb{E}[\mathbf{1}_A|\mathcal{G}]$$

Remark B.66. It is important to distinguish between the probability of an event given another event (Definition B.26), which is a true probability measure, and the probability of an event given a σ-algebra (Definition B.65), which is a random variable.

By Definition B.57 we have that:

i) $\mathbb{P}(A|\mathcal{G})$ is $\mathcal{G}$-measurable and integrable.

ii) $\int_G \mathbb{P}(A|\mathcal{G})\, d\mathbb{P} = \mathbb{P}(A \cap G) \ \forall G \in \mathcal{G}$

where in the last equality we have used that

$$\int_G \mathbf{1}_A d\mathbb{P} = \int_\Omega \mathbf{1}_A \mathbf{1}_G d\mathbb{P} = \int_{A\cap G} d\mathbb{P}$$

According to the previous interpretation of conditional expectations, $\mathbb{P}(A|\mathcal{G})(\omega)$ can be understood as the probability of A for someone who knows for each G in $\mathcal{G}$ whether or not it contains the point ω.

In order to obtain the specific relationship between $\mathbb{P}(A|B)$ and $\mathbb{P}[A|\sigma(B)]$, with B an event, and $\sigma(B)$ the σ-algebra generated by B, we need an additional result.

Proposition B.67. *Let $\{\Lambda_n, n \geq 1\}$ be a collection of events with $\mathbb{P}(\Lambda_n) > 0 \ \forall n \geq 1$, and such that $\Lambda_i \cap \Lambda_j = \varnothing$ for $i \neq j$ and $\cup_n \Lambda_n = \Omega$, i.e., $\{\Lambda_n, n \geq 1\}$ is a partition of Ω. If X is an integrable random variable and A is an event, then* *i)*

$$\mathbb{E}\left[X\,|\sigma\left(\{\Lambda_n, n \geq 1\}\right)\right] = \sum_{n\geq 1} \mathbb{E}\left[X\,|\Lambda_n\right] \mathbf{1}_{\Lambda_n} \ a.s.$$

ii)

$$\mathbb{P}\left[A\,|\{\Lambda_n, n \geq 1\}\right] = \sum_{n\geq 1} \mathbb{P}\left(A\,|\Lambda_n\right) \mathbf{1}_{\Lambda_n} \ a.s.$$

Proof. See Resnick (2014), Example 10.2.1. □

The following corollary states the relationship between the conditional probabilities $\mathbb{P}(A\,|B)$ and $\mathbb{P}[A\,|\sigma(B)]$.

Corollary B.68. *Let A, B be two events with $\mathbb{P}(B), \mathbb{P}(B^c) > 0$, where B^c is the complement of B, $B^c = \Omega - B$. Then*

$$\mathbb{P}[A\,|\sigma(B)] = \mathbb{P}(A\,|B)\,\mathbf{1}_B + \mathbb{P}\left(A\,\middle|B^c\right)\mathbf{1}_{B^c} \ a.s.$$

Proof. We have that $\sigma(B) = \{\varnothing, B, B^c, \Omega\} = \sigma[\{B, B^c\}]$. As $\{B, B^c\}$ is a partition of Ω, applying Proposition B.67 we obtain the desired result. □

The main properties of conditional probabilities with respect to a σ-algebra are summarized in the following result.

Proposition B.69. *Let A, B be events, $\{A_n\}_{n\geq 1}$ a finite or countable sequence of events, and $\mathcal{G}$ a sub-σ-algebra. Then the following properties hold:*

i) $\mathbb{P}(\varnothing\,|\mathcal{G}) = 0$ *and* $\mathbb{P}(\Omega\,|\mathcal{G}) = 1$ *a.s.*

ii) *If* $\mathbb{P}(A) = 1$, *then* $\mathbb{P}(A\,|\mathcal{G}) = 1$ *a.s.*

iii) *If* $\mathbb{P}(A) = 0$, *then* $\mathbb{P}(A\,|\mathcal{G}) = 0$ *a.s.*

iv) $0 \leq \mathbb{P}(A\,|\mathcal{G}) \leq 1$ *a.s.*

v) *If A is $\mathcal{G}$-measurable, then* $\mathbb{P}(A\,|\mathcal{G}) = \mathbf{1}_A$ *a.s.*

vi) $\mathbb{P}[A\,|\{\varnothing, \Omega\}] = \mathbb{P}(A)$ *a.s.*

vii) *A is independent of $\mathcal{G}$ if, and only if,* $\mathbb{P}(A\,|\mathcal{G}) = \mathbb{P}(A)$ *a.s.*

viii) *If $A_i \cap A_j = \varnothing$ for $i \neq j$, then* $\mathbb{P}\left(\bigcup_{n\geq 1} A_n\,|\mathcal{G}\right) = \sum_{n\geq 1} \mathbb{P}(A_n\,|\mathcal{G})$ *a.s.*

ix) *If $A \subseteq B$, then, with probability one,* $\mathbb{P}(B - A\,|\mathcal{G}) = \mathbb{P}(B\,|\mathcal{G}) - \mathbb{P}(A\,|\mathcal{G})$ *and* $\mathbb{P}(A\,|\mathcal{G}) \leq \mathbb{P}(B\,|\mathcal{G})$ *a.s.*

Proof. See Billingsley (1995), Section 33, pages 431-437. □

The rest of this section is dedicated to defining and studying conditional expectations with respect to random variables and particular values of them.

Definition B.70. (Conditional expectation with respect to a random variable) Let X, Y be two random variables with $X \in L^1(\Omega, \mathcal{F}, \mathbb{P})$. The conditional expectation of X with respect to Y, $\mathbb{E}[X\,|Y]$, is the random variable defined as

$$\mathbb{E}[X\,|Y] = \mathbb{E}[X\,|\sigma(Y)]$$

The definition of conditional probability, given a random variable, follows naturally from the previous definition.

Definition B.71. (Conditional probability given a random variable) Let A be an event, and Y a random variable. Then, the conditional probability of A given Y, $\mathbb{P}(A|Y)$, is defined as

$$\mathbb{P}(A|Y) = \mathbb{E}\left[\mathbf{1}_A|Y\right]$$

The above definitions can be extended to conditional expectations and probabilities with respect to a collection of random variables. Moreover, all the previous results related to conditional expectations with respect to σ-algebras can be applied to them.

When defining conditional expectations given a specific value of a random variable, $Y = y$, one is tempted to use expression (B.6) for the conditional expectation $\mathbb{E}\left[X\,|Y = y\right]$, but it is not defined in the continuous case, since $\mathbb{P}\left[Y = y\right] = 0$. In what follows, we provide an alternative path.

Theorem B.72. *Let X, Y be two random variables with $X \in L^1(\Omega, \mathcal{F}, \mathbb{P})$. Then, there exists a Borel measurable function $\varphi : \mathbb{R} \to \mathbb{R}$ such that $\mathbb{E}\left[X\,|Y\right] = \varphi(Y)$ a.s.*

Proof. See Chung (2001), Theorem 9.1.2. □

Definition B.73. (Conditional expectation with respect to a particular value of a random variable) Let X, Y be two random variables with $X \in L^1(\Omega, \mathcal{F}, \mathbb{P})$. The conditional expectation of X with respect to $Y = y$, $\mathbb{E}\left[X\,|Y = y\right]$, is defined by

$$\mathbb{E}\left[X\,|Y = y\right] = \varphi(y)$$

B.7 Multidimensional Distributions

In this section we study random variables with values in $\mathbb{R}^n$, with the main objective of applying this knowledge to the multivariate normal distribution. First, we need to extend the concept of measurability to $\mathbb{R}^n$-valued functions.

Definition B.74. (Measurable vector function) Let $(X, \mathcal{F}, \mu)$ be a measure space. A function $f : X \to \mathbb{R}^n$ is $\mathcal{F}$-measurable if $f^{-1}(B) \in \mathcal{F}$ for all $B \in \mathcal{B}(\mathbb{R}^n)$, the Borel σ-algebra on $\mathbb{R}^n$.

Definition B.75. (Random vector) Let $(\Omega, \mathcal{F}, \mathbb{P})$ be a probability space. A random vector is an $\mathcal{F}$-measurable function from Ω to $\mathbb{R}^n$.

From the definition, we have that any random n-vector $\mathbf{X}$ must be of the form

$$\mathbf{X}(\omega) = \left(X_1(\omega), \ldots, X_n(\omega)\right)^\top \quad \omega \in \Omega$$

Proposition B.76. $\mathbf{X} = (X_1, \ldots, X_n)^\top$ *is a random vector if, and only if, all its components $X_1, \ldots, X_n$ are random variables.*

Proof. See Billingsley (1995), page 255. □

According to the previous proposition, all random vectors $\mathbf{X} : \Omega \to \mathbb{R}^n$ are composed of n random variables. Conversely, any vector of random variables is also a random vector.

Definition B.77. (σ-algebra generated by a random vector) The σ-algebra generated by the random vector $\mathbf{X}$, $\sigma(\mathbf{X})$, is the smallest σ-algebra with respect to which $\mathbf{X}$ is measurable.

Definition B.78. (Independence of random vectors) Let $\mathbf{X}, \mathbf{Y}$ be two random vectors. We say that $\mathbf{X}$ and $\mathbf{Y}$ are independent if $\sigma(\mathbf{X})$ and $\sigma(\mathbf{Y})$ are independent σ-algebras.

Definition B.79. (Expectation and covariance of a random vector) If $\mathbf{X} = (X_1, \ldots, X_n)^\top$ is a random vector, then its expectation is the n-vector given by

$$\mathbb{E}[\mathbf{X}] = \int_\Omega \mathbf{X}(\omega)\, d\mathbb{P}(\omega) = \left(\int_\Omega X_1(\omega)\, d\mathbb{P}(\omega), \ldots, \int_\Omega X_n(\omega)\, d\mathbb{P}(\omega) \right)^\top$$

and its covariance matrix is the matrix $\mathbf{\Sigma} = (\Sigma_{ij})$, with elements $\Sigma_{ij} = \text{Cov}(X_i, X_j)$, $i, j = 1, \ldots, n$.

Proposition B.80. *Let $\mathbf{X} = (X_1, \ldots, X_n)^\top$ be a random vector with $X_i \in L^2(\Omega, \mathcal{F}, \mathbb{P})$ $\forall i \in \{1, \ldots, n\}$, and let $\mathbf{\Sigma}$ be its covariance matrix. Then, the following properties hold:*

i) $\mathbf{\Sigma}$ *is symmetric.*

ii) $\Sigma_{ii} = \text{Var}(X_i)$

iii) If X_i and X_j are independent for $i \neq j$, then $\mathbf{\Sigma}$ is diagonal.

iv) $\mathbf{\Sigma}$ *is positive semidefinite.*

Proof. Properties i)-iii) are immediate from the properties of the covariance of two random vectors (see Propositions B.23 and B.36). For property iv), it is enough to realize that $\mathbf{a}^\top \mathbf{\Sigma} \mathbf{a} = \text{Var}(\mathbf{a}^\top \mathbf{X}) \geq 0$ $\forall \mathbf{a} \in \mathbb{R}^n$. □

The following result is the equivalent to Proposition B.7 for random vectors.

Proposition B.81. *Let $(\Omega, \mathcal{F}, \mathbb{P})$ be a probability space, and $\mathbf{X}$ an $\mathbb{R}^n$-valued random vector on $(\Omega, \mathcal{F})$. Then, the mapping $\mathbb{P}_\mathbf{X} : \mathcal{B}(\mathbb{R}^n) \to [0, 1]$ defined by*

$$\mathbb{P}_\mathbf{X}(B) = \mathbb{P}(\mathbf{X} \in B) = \mathbb{P}\left[\mathbf{X}^{-1}(B)\right]$$

is a probability on $(\mathbb{R}^n, \mathcal{B}(\mathbb{R}^n))$. Moreover, if $g : \mathbb{R}^n \to \mathbb{R}$ is a Borel measurable function, then

$$g(\mathbf{X}) \in L^1(\Omega, \mathcal{F}, \mathbb{P}) \Leftrightarrow g \in L^1\left(\mathbb{R}^n, \mathcal{B}(\mathbb{R}^n), \mathbb{P}_\mathbf{X}\right)$$

and if these function are integrable

$$\int_\Omega [g(\mathbf{X}(\omega))]\, d\mathbb{P}(\omega) = \int_{\mathbb{R}^n} g(\mathbf{x})\, d\mathbb{P}_\mathbf{X}(\mathbf{x})$$

Proof. It suffices to realize that $\mathbb{P}_\mathbf{X} = \mathbb{P} \circ \mathbf{X}^{-1}$ and apply Proposition 2.6.8 of Cohn (2013), which is the general version of Proposition A.27. □

Definition B.82. (Probability distribution and distribution function of a random vector) Let $(\Omega, \mathcal{F}, \mathbb{P})$ be a probability space, and $\mathbf{X} : \Omega \to \mathbb{R}^n$ a random vector on $(\Omega, \mathcal{F})$. The probability $\mathbb{P}_{\mathbf{X}} : \mathcal{B}(\mathbb{R}^n) \to [0,1]$ defined in Proposition B.81 is called the probability distribution of $\mathbf{X}$. The function $F_{\mathbf{X}} : \mathbb{R}^n \to \mathbb{R}$ defined for each $\mathbf{x} = (x_1, \ldots, x_n)^\top \in \mathbb{R}^n$ by

$$F_{\mathbf{X}}(\mathbf{x}) = \mathbb{P}_{\mathbf{X}}\left(\left\{\mathbf{y} \in \mathbb{R}^n : y_i \le x_i,\ i = 1, \ldots, n\right\}\right) = \mathbb{P}(X_1 \le x_1, \ldots, X_n \le x_n) \quad \text{(B.7)}$$

is the distribution function of $\mathbf{X}$.

Definition B.83. (Joint distribution function) If $X_1, \ldots, X_n$ are random variables, the distribution function of the random vector $(X_1, \ldots, X_n)^\top$ is called the joint distribution function of $X_1, \ldots, X_n$.

Proposition B.84. *The distribution function of the random vector* $\mathbf{X} = (X_1, \ldots, X_n)^\top$ *satisfies the following properties:*

i) $F_{\mathbf{X}}$ *is increasing in each variable.*

ii) $\lim_{h \to 0^+} F_{\mathbf{X}}(x_1 + h, \ldots, x_n + h) = F(x_1, \ldots, x_n)$

iii) $\lim_{x_1, \ldots, x_n \to +\infty} F_{\mathbf{X}}(x_1, \ldots, x_n) = 1$

iv) $\lim_{x_i \to -\infty} F_{\mathbf{X}}(x_1, \ldots, x_n) = 0 \ \ \forall i \in \{1, \ldots, n\}$

Proof. See Billingsley (1995), page 260. □

Definition B.85. (Density function of a random vector) Let $\mathbf{X}$ be a random vector, and $F_{\mathbf{X}}$ its distribution function. If there exists a non-negative and Lebesgue integrable function $f_{\mathbf{X}}$:$\mathbb{R}^n \to \mathbb{R}$ such that

$$\mathbb{P}_{\mathbf{X}}(A) = \int_A f_{\mathbf{X}}(\mathbf{x})\, d\mathbf{x} \quad \forall A \in \mathcal{B}\left(\mathbb{R}^n\right) \qquad \text{(B.8)}$$

we say that $f_{\mathbf{X}}$ is the density function of $\mathbf{X}$.

Proposition B.86. *Let* $\mathbf{X}$ *be a random vector with density* $f_{\mathbf{X}}$*, and let* $F_{\mathbf{X}}$ *be its distribution function. Then*

$$F_{\mathbf{X}}(x_1, \ldots, x_n) = \int_{-\infty}^{x_1} \cdots \int_{-\infty}^{x_n} f_{\mathbf{X}}(y_1, \ldots, y_n)\, dy_n \cdots dy_1$$

Proof. An immediate consequence of expressions (B.7) and (B.8). □

Proposition B.87. *Let* $\mathbf{X}$ *be a random vector with density* $f_{\mathbf{X}}$*, and let* $g : \mathbb{R}^n \to \mathbb{R}$ *be a function such that* $g(\mathbf{X}) \in L^1(\Omega, \mathcal{F}, \mathbb{P})$*. Then*

$$\mathbb{E}[g(\mathbf{X})] = \int_{\mathbb{R}^n} g(\mathbf{x})\, f_{\mathbf{X}}(\mathbf{x})\, d\mathbf{x}$$

Proof. Analogous to the proof of Proposition B.17. □

Definition B.88. (Marginal probability distribution, marginal distribution function) Let $\mathbf{X} = (X_1, \ldots, X_n)$ be a random vector on the probability space $(\Omega, \mathcal{F}, \mathbb{P})$. The mapping $\mathbb{P}_{X_i} : \mathcal{B}(\mathbb{R}) \longrightarrow [0, 1]$ given for each $i \in \{1, \ldots, n\}$ by

$$\mathbb{P}_{X_i}(A) = \mathbb{P}_{\mathbf{X}}[(x_1, \ldots, x_n) : x_i \in A] = \mathbb{P}[X_i \in A]$$

is called the marginal probability distribution of X_i, and coincides with the probability distribution of the random variable X_i. The function $F_{X_i} : \mathbb{R} \to \mathbb{R}$ defined by

$$F_{X_i}(x) = \mathbb{P}_{X_i}((-\infty, x]) = \mathbb{P}[X_i \leq x]$$

is the marginal distribution function of X_i, and coincides with the distribution function of the random variable X_i.

Proposition B.89. *If* $\mathbf{X}$ *has density* $f_{\mathbf{X}}$*, then the density function associated to* F_{X_i}*, which coincides with the density function of the random variable* X_i*, can be obtained as*

$$f_{X_i}(x) = \int_{\mathbb{R}^{n-1}} f_{\mathbf{X}}(x_1, \ldots, x_{i-1}, x, x_{i+1}, \ldots, x_n)\, dx_1 \cdots dx_{i-1} dx_{i+1} \cdots dx_n$$

Proof. See Billingsley (1995), page 261. □

Proposition B.90. *Let* $\mathbf{X} = (X_1, \ldots, X_n)$ *be a random vector. Then, the following statements are equivalent:*

i) The random variables $X_1, \ldots, X_n$ *are independent.*

ii)

$$\mathbb{P}(X_1 \in B_1, \ldots, X_n \in B_n) = \mathbb{P}(X_1 \in B_1) \cdots \mathbb{P}(X_n \in B_n) \ \ \forall B_1, \ldots, B_n \in \mathcal{B}(\mathbb{R})$$

iii)

$$F_{\mathbf{X}}(x_1, \ldots, x_n) = F_{X_1}(x_1) \cdots F_{X_n}(x_n) \ \ \forall x_1, \ldots, x_n \in \mathbb{R}$$

Moreover, if the random variables have densities, then they are independent if, and only if

$$f_{\mathbf{X}}(x_1, \ldots, x_n) = f_{X_1}(x_1) \cdots f_{X_n}(x_n) \ \ \forall x_1, \ldots, x_n \in \mathbb{R}$$

Proof. See Billingsley (1995), pages 261-262. □

Definition B.91. (Characteristic function of a random vector) Let $\mathbf{X} = (X_1, \ldots, X_n)$ be a random vector. The characteristic function of $\mathbf{X}$ is the function $\varphi_{\mathbf{X}} : \mathbb{R}^n \to \mathbb{C}$ defined by

$$\varphi_{\mathbf{X}}(\mathbf{u}) = \mathbb{E}\left[e^{i\langle \mathbf{u}, \mathbf{X}\rangle}\right]$$

As in the one-dimensional case, we have that $|\varphi_{\mathbf{X}}(\mathbf{u})| \leq 1 \ \forall \mathbf{u} \in \mathbb{R}^n$, and then, it is well defined and finite for every random vector.

Proposition B.92. *Let* $\mathbf{X}, \mathbf{Y}$ *be two random n-vectors, and* $\varphi_{\mathbf{X}}, \varphi_{\mathbf{Y}}$ *their respective characteristic functions. Then:*

i) $\varphi_{\mathbf{X}}(\mathbf{0}) = 1$

ii) If $\mathbf{Y} = \mathbf{AX} + \mathbf{b}$*, with* $\mathbf{A} \in \mathbb{R}^{k \times n}$ *and* $\mathbf{b} \in \mathbb{R}^k$*, then*

$$\varphi_{\mathbf{Y}}(\mathbf{u}) = e^{i\langle \mathbf{u}, \mathbf{b}\rangle} \varphi_{\mathbf{X}}\left(\mathbf{A}^\top \mathbf{u}\right) \quad \forall \mathbf{u} \in \mathbb{R}^k$$

iii) If $\mathbf{X}$ *and* $\mathbf{Y}$ *are independent, then*

$$\varphi_{\mathbf{X}+\mathbf{Y}}(\mathbf{u}) = \varphi_{\mathbf{X}}(\mathbf{u})\, \varphi_{\mathbf{Y}}(\mathbf{u}) \quad \forall \mathbf{u} \in \mathbb{R}^n$$

iv) $\varphi_{\mathbf{X}}$ *is uniformly continuous on* $\mathbb{R}^n$*.*

v) If $\varphi_{\mathbf{X}}(\mathbf{u}) = \varphi_{\mathbf{Y}}(\mathbf{u})$ *for all* $\mathbf{u} \in \mathbb{R}^n$*, then* $F_{\mathbf{X}}(\mathbf{x}) = F_{\mathbf{Y}}(\mathbf{x}) \quad \forall \mathbf{x} \in \mathbb{R}^n$

vi) If $\mathbf{X} = (X_1, \ldots, X_n)$*, then* $X_1, \ldots, X_n$ *are independent if, and only if,*

$$\varphi_{\mathbf{X}}(\mathbf{u}) = \varphi_{X_1}(u_1) \cdots \varphi_{X_n}(u_n) \quad \forall \mathbf{u} = (u_1, \ldots, u_n) \in \mathbb{R}^n$$

vii) If $\varphi_{\mathbf{X}}$ *is integrable, then* $\mathbf{X}$ *has a bounded continuous density* $f_{\mathbf{X}}$*, given by*

$$f_{\mathbf{X}}(\mathbf{x}) = \frac{1}{(2\pi)^n} \int_{\mathbb{R}^n} e^{-i\langle \mathbf{y}, \mathbf{x}\rangle} \varphi_{\mathbf{X}}(\mathbf{y})\, d\mathbf{y}$$

Proof. Property i) is immediate. For the remaining properties the references are the following: ii) Jacod and Protter (2004) Theorem 13.3, iii) Evans (2013), page 23, iv) Resnick (2014), page 296, v) Jacod and Protter (2004), Theorem 14.1, vi) Cohn (2013), Corollary 10.3.12, vii) Cramer (1946), page 101. □

B.8 The Multivariate Normal Distribution

In this section we illustrate the results related to multidimensional distributions with the most important example of this type of distributions: the multivariate normal.

Definition B.93. (Multivariate normal distribution) The random vector determined by $\mathbf{X} = (X_1, \ldots, X_n)^\top$ has a multivariate normal distribution (or Gaussian distribution) if every linear combination $\sum_{i=1}^{n} a_i X_i$, with $a_i \in \mathbb{R}$, has a one-dimensional normal distribution.

Remark B.94. In the previous definition we have to consider the so-called degenerate one-dimensional normal distribution, $N(0, 0)$, corresponding to the case $a_1 = \cdots = a_n = 0$.

The basic properties of the multivariate normal distribution appear in the following results.

Proposition B.95. *Let* $\mathbf{X} = (X_1, \ldots, X_n)$ *be a multivariate normal random vector. Then, the characteristic function of* $\mathbf{X}$ *is of the form*

$$\varphi_{\mathbf{X}}(\mathbf{u}) = \exp\left\{ i\mathbf{u}^\top \boldsymbol{\mu} - \frac{1}{2}\mathbf{u}^\top \boldsymbol{\Sigma} \mathbf{u} \right\}$$

where $\boldsymbol{\mu}$ *and* $\boldsymbol{\Sigma}$ *are, respectively, the expectation and the covariance matrix of the random vector* $\mathbf{X}$*.*

Proof. See Jacod and Protter (2004), Theorem 16.1. □

According to the previous proposition, the distribution of a multivariate normal random vector is completely determined by the expectation or mean vector, $\boldsymbol{\mu} = \mathbb{E}[\mathbf{X}]$, and the covariance matrix, with elements $\Sigma_{ij} = \operatorname{Cov}(X_i, X_j)$. If $\mathbf{X} = (X_1, \ldots, X_n)$ has a multivariate normal distribution with expectation $\boldsymbol{\mu}$ and covariance matrix $\boldsymbol{\Sigma}$, we write $\mathbf{X} \sim N(\boldsymbol{\mu}, \boldsymbol{\Sigma})$.

The following result, which is a direct consequence of the previous proposition, states that the family of multivariate normal distributions is closed under affine transformations.

Theorem B.96. *Let $\mathbf{X} \sim N(\boldsymbol{\mu}, \boldsymbol{\Sigma})$ be a normal random n-vector, and let $\mathbf{A} \in \mathbb{R}^{m\times n}$ and $\mathbf{b} \in \mathbb{R}^m$. Then the random vector $\mathbf{Y}$, defined by the transformation $\mathbf{Y} = \mathbf{AX} + \mathbf{b}$ is also Gaussian and*

$$\mathbf{Y} \sim N\left(\mathbf{A}\mu + \mathbf{b}, \mathbf{A}\boldsymbol{\Sigma}\mathbf{A}^\top\right).$$

Proof. Since the characteristic function of $\mathbf{X} \sim N(\boldsymbol{\mu}, \boldsymbol{\Sigma})$ is

$$\varphi_{\mathbf{X}}(\mathbf{u}) = \exp\left(i\mathbf{u}^\top \mu - \frac{1}{2}\mathbf{u}^\top \boldsymbol{\Sigma}\mathbf{u}\right)$$

we can write the characteristic function of $\mathbf{Y}$ as

$$\begin{aligned}\varphi_{\mathbf{Y}}(\mathbf{u}) &= \mathbb{E}\left[e^{i\mathbf{u}^\top(\mathbf{AX}+\mathbf{b})}\right] = e^{i\mathbf{u}^\top\mathbf{b}}\mathbb{E}\left[e^{i(\mathbf{A}^\top\mathbf{u})^\top\mathbf{X}}\right] = e^{i\mathbf{u}^\top\mathbf{b}}\varphi_{\mathbf{X}}(\mathbf{A}^\top\mathbf{u})\\ &= \exp\left(i\mathbf{u}^\top(\mathbf{A}\mu + \mathbf{b}) - \frac{1}{2}\mathbf{u}^\top\mathbf{A}\boldsymbol{\Sigma}\mathbf{A}^\top\mathbf{u}\right)\end{aligned}$$

which concludes the proof. □

The joint distribution function of a finite number of independent and normal random variables is a multivariate normal random variable.

Proposition B.97. *Let $X_1, \ldots, X_n$ be independent random variables, with $X_i \sim N\left(\mu_i, \sigma_i^2\right)$, $i = 1, \ldots, n$. Then, $\mathbf{X} = (X_1, \ldots, X_n)$ is a random vector with a multivariate normal distribution $N(\boldsymbol{\mu}, \boldsymbol{\Sigma})$, where $\boldsymbol{\mu} = (\mu_1, \ldots, \mu_n)$ and $\boldsymbol{\Sigma} = \operatorname{diag}\left(\sigma_1^2, \ldots, \sigma_n^2\right)$.*

Proof. See Jacod and Protter (2004), Example 1, page 127. □

Remark B.98. The independence assumption is essential for the proof of the previous proposition. In general, a collection of one-dimensional normal random variables need not have a multivariate normal joint distribution.

By Proposition B.97, if $X_1, \ldots, X_n$ are independent normal random variables, then $\mathbf{X} = (X_1, \ldots, X_n)$ is multivariate normal with a diagonal covariance matrix. The converse is also true.

Proposition B.99. *Let $\mathbf{X} = (X_1, \ldots, X_n)$ be a random vector with multivariate normal distribution. Then, if its covariance matrix is diagonal, the random variables $X_1, \ldots, X_n$ are independent.*

Proof. See Jacod and Protter (2004), Corollary 16.1. □

Remark B.100. From ii) of Proposition B.36, we know that the independence of random variables implies that these variables are uncorrelated, that is, their covariance is zero. By the previous proposition, for multivariate normal random variables, the converse is also true. In other words, for random variables with a multivariate normal joint distribution, being uncorrelated is equivalent to independence.

The following theorem states that every random n-vector with multivariate normal distribution can be obtained by a linear transformation of vectors of independent random variables with one-dimensional normal distributions.

Theorem B.101. *Let* $\mathbf{X}$ *be a random vector with multivariate normal distribution and* $\boldsymbol{\mu} = \mathbb{E}[\mathbf{X}]$. *Then, there exist independent normal random variables,* $Y_1, \ldots, Y_n$, *such that* $Y_i \sim N\left(0, \sigma_i^2\right)$, $i = 1, \ldots, n$, *and an orthogonal matrix* $\mathbf{A}$, *such that* $\mathbf{X} = \boldsymbol{\mu} + \mathbf{A}\mathbf{Y}$, *with* $\mathbf{Y} = (Y_1, \ldots, Y_n)^\top$.

Proof. See Jacod and Protter (2004), Theorem 16.2. □

Corollary B.102. *Let* $\mathbf{X} = (X_1, \ldots, X_n)$ *be a random vector with* $\mathbf{X} \sim N(\boldsymbol{\mu}, \boldsymbol{\Sigma})$. *Then,* $\mathbf{X}$ *has a density if, and only if,* $|\boldsymbol{\Sigma}| \neq 0$. *Moreover, if* $|\boldsymbol{\Sigma}| \neq 0$, *the density function of* $\mathbf{X}$ *is given by*

$$f_{\mathbf{X}}(\mathbf{x}) = \frac{1}{(2\pi)^{n/2}\sqrt{|\boldsymbol{\Sigma}|}} \exp\left\{-\frac{1}{2}(\mathbf{x} - \boldsymbol{\mu})^\top \boldsymbol{\Sigma}^{-1} (\mathbf{x} - \boldsymbol{\mu})\right\}$$

Proof. See Jacod and Protter (2004), Corollary 16.2. □

Bibliography

Axler, S. (2020). *Measure, Integration & Real Analysis*. Springer.

Bass, R. F. (2011). *Stochastic Processes*. Cambridge University Press.

Billingsley, P. (1995). *Probability and Measure*. Wiley-Interscience, third edition.

Bjork, T. and Christensen, B. J. (1999). Interest rate dynamics and consistent forward rate curves. *Mathematical Finance*, 9(4):323–348.

Björk, T. (2020). *Arbitrage Theory in Continuous Time*. Oxford University Press, fourth edition.

Bueno-Guerrero, A. (2020). *The Mathematics of Financial Markets Within Everyone's Reach: An Introduction Without Formulas in the Main Text*. Self-published.

Bueno-Guerrero, A., Moreno, M., and Navas, J. F. (2015). Stochastic string models with continuous semimartingales. *Physica A: Statistical Mechanics and its Applications*, 433:229–246.

Bueno-Guerrero, A., Moreno, M., and Navas, J. F. (2016). The stochastic string model as a unifying theory of the term structure of interest rates. *Physica A: Statistical Mechanics and its Applications*, 461:217–237.

Calin, O. (2015). *An Informal Introduction to Stochastic Calculus with Applications*. World Scientific.

Chung, K. L. (2001). *A Course in Probability Theory*. Academic Press, third edition.

Cohn, D. L. (2013). *Measure Theory*. Birkhäuser.

Cramer, H. (1946). *Mathematical Methods of Statistics*. Asia Publishing House.

Durrett, R. (1996). *Stochastic Calculus: A Practical Introduction*. CRC Press.

Einstein, A. (1956). *Investigations on the Theory of the Brownian Movement*. Dover.

Evans, L. C. (2013). *An Introduction to Stochastic Differential Equations*. American Mathematical Society.

Falconer, K. (2003). *Fractal Geometry: Mathematical Foundations and Applications*. Wiley, second edition.

Feynman, R. (1948). Space-time approach to non-relativistic quantum mechanics. *Review of Modern Physics*, 20(2):367–387.

Folland, G. B. (1999). *Real Analysis: Modern Techniques and Their Applications*. John Wiley & Sons, second edition.

Friedman, A. (2006). *Stochastic Differential Equations and Applications*. Dover Publications.

Halmos, P. R. (1950). *Measure Theory*. Springer-Verlag.

Haussermann, J. W. (1989). An introduction to fractal and chaos. Technical report, Defense Technical Information Center.

Itô, K. (1944). Stochastic integral. *Proceedings of the Imperial Academy, Tokio*, 20:519–524.

Jacod, J. and Protter, P. (2004). *Probability Essentials*. Springer.

Jordan, P. and Neumann, J. (1935). On inner products in linear metric spaces. *Annals of Mathematics*, 36(3):719–723.

Kac, M. (1949). On distributions of certain wiener functionals. *Transactions of the American Mathematical Society*, 65(1):1–13.

Kallenberg, O. (2021). *Foundations of Modern Probability*. Springer, third edition.

Kallianpur, G. and Sundar, P. (2014). *Stochastic Analysis and Diffusion Processes*. Oxford University Press.

Karatzas, I. and Shreve, S. E. (1991). *Brownian Motion and Stochastic Calculus*. Springer, second edition.

Kechris, A. S. (1995). *Classical Descriptive Set Theory*. Springer-Verlag.

Klebaner, F. C. (2005). *Introduction to Stochastic Calculus with Applications*. Imperial College Press, second edition.

Kloeden, P. E. and Platen, E. (1992). *Numerical Solution of Stochastic Differential Equations*. Springer.

Kolokoltsov, V. N. (2011). *Markov Processes, Semigroups and Generators*. De Gruyter.

Kuo, H.-H. (2006). *Introduction to Stochastic Integration*. Springer.

Le Gall, J.-F. (2016). *Brownian Motion, Martingales, and Stochastic Calculus*. Springer.

Lipster, R. S. and Shiryaev, A. N. (2001). *Statistic of Random Processes I: General Theory*. Springer, second edition.

Mandelbrot, B. B. (1982). *The Fractal Geometry of Nature*. W.H. Freeman and Company.

Mao, X. (2008). *Stochastic Differential Equations and Applications*. Woodhead Publishing, second edition.

McKean, H.P., J. (1969). *Stochastic Integrals*. Academic Press.

Mörters, P. and Peres, Y. (2010). *Brownian Motion*. Cambridge University Press.

Nualart, D. (2006). *The Malliavin Calculus and Related Topics*. Springer, second edition.

Oksendal, B. (1998). *Stochastic Differential Equations: An Introduction with Applications*. Springer, fifth edition.

Owen, D. B. (1980). A table of normal integrals. *Communications in Statistics - Simulation and Computation*, 4(9):389–419.

Paley, R., Wiener, N., and Zygmund, A. (1933). Notes on random functions. *Mathematische Zeitschrift*, 1(37):647–668.

Pozniak, A. S. (2009). *Advanced Mathematical Tools for Automatic Control Engineers. Volume 2: Stochastic Techniques*. Elsevier.

Protter, P. E. (2005). *Stochastic Integration and Differential Equations*. Springer, second edition.

Resnick, S. I. (2014). *A Probability Path*. Springer.

Revuz, D. and Yor, M. (1999). *Continuous Martingales and Brownian Motion*. Springer, third edition.

Rockafellar, R. T. (1970). *Convex Analysis*. Princeton University Press.

Rogers, L. and Williams, D. (1994). *Diffusions, Markov Processes, and Martingales. Volume I: Foundations*. John Wiley & Sons, second edition.

Royden, H. and Fitzpatrick, P. (2010). *Real Analysis*. Prentice Hall, fourth edition.

Rudin, W. (1987). *Real and Complex Analysis*. McGraw-Hill, third edition.

Santa-Clara, P. and Sornette, D. (2001). The dynamics of the forward interest rate curve with stochastic string shocks. *Review of Financial Studies*, 14(1):149–185.

Schilling, R. L. and Partzsch, L. (2012). *Brownian Motion: An Introduction to Stochastic Processes*. De Gruyter.

Shreve, S. E. (2004). *Stochastic Calculus for Finance II: Continous-Time Models*. Springer.

Steele, J. M. (2001). *Stochastic Calculus and Financial Applications*. Springer.

Toral, R. and Colet, P. (2014). *Stochastic Numerical Methods: An Introduction for Students and Scientists*. Wiley-VCH.

Wiener, N. (1923). Differential space. *Journal of Mathematical Physics*.

Index